Nurse Management Demystified

Irene McEachen, RN, MSN, Ed.D.
Jim Keogh

New York   Chicago   San Francisco   Lisbon   London
Madrid   Mexico City   Milan   New Delhi   San Juan
Seoul   Singapore   Sydney   Toronto

The **McGraw·Hill** Companies

McGraw-Hill books are available at special quantity discounts to use as premiums and sales promotions, or for use in corporate training programs. For more information, please write to the Director of Special Sales, Professional Publishing, McGraw-Hill, Two Penn Plaza, New York, NY 10121-2298. Or contact your local bookstore.

Nurse Management Demystified

1234567890 DOC DOC 019876

ISBN-13:   978-0-07-147241-8
ISBN-10:   0-07-147241-X

Sponsoring Editor
Judy Bass

Editorial Supervisor
Janet Walden

Project Manager
Vastavikta Sharma

Copy Editor
William McManus

Proofreader
Ben Weiler

Indexer
Robert Swanson

Production Supervisor
Jean Bodeaux

Composition
International Typesetting
and Composition

Illustration
International Typesetting
and Composition

Cover Series Design
Margaret Webster-Shapiro

Cover Illustration
Lance Lekander

To all the current and future Nurse Managers...
important members of the healthcare team.
—Irene McEachen

This book is dedicated to Anne, Sandy, Joanne,
Amber-Leigh Christine, and Graaf, without whose
help and support this book couldn't have been
written.
—Jim Keogh

ABOUT THE AUTHORS

Irene McEachen, R.N., Ed.D. was formerly vice president of nursing at Beth Israel Medical Center in New York City. She is currently a professor of nursing at Saint Peter's College in New Jersey, where she teaches undergraduate and graduate courses in nursing administration.

Jim Keogh was formerly on the faculty of Columbia University and is on the faculty of New York University and Saint Peter's College. He is the coauthor of *Microbiology Demystified* and *Pharmacology Demystified* as well as several other books.

The authors acknowledge, with deep appreciation, the thoughtful and insightful input they received from Helen Reyes, RN, MSN, for her careful review of this manuscript.

CONTENTS

INTRODUCTION

Caring for up to seven patients is challenging for any registered nurse, but how would you feel if you were responsible for the healthcare of 50 patients, 7 days a week, 24 hours a day—and for managing a staff of 40? Just the thought is enough to send shivers down the spine of even the most battled-scarred nurse. Yet, these are just some of a Nurse Manager's responsibilities when supervising the day-to-day operations of a clinical unit in an average healthcare facility.

A Nurse Manager is in charge of patient care, staffing, scheduling, budgeting, planning—everything that is necessary to keep a busy clinical unit financially viable while delivering high-quality healthcare to every patient on the floor.

Patients look to their nurse to treat their healthcare problems. Nurses look to the Nurse Manager to help solve their problems, whether it is handling a difficult patient, dealing with an uncooperative physician, or more personal matters such as training, shift assignments, and career advancement.

Hospital administrators expect Nurse Managers to be the jack of all trades and master of running daily operations of a unit within the budget and policies approved by the hospital's board of trustees.

The Nurse Manager must do the following:

- Thoroughly understand healthcare economics
- Lead teams of healthcare professionals and support staff
- Anticipate needs and develop a strategy to fulfill those needs
- Know how to delegate and supervise a large staff
- Use effective communications to resolve conflicts
- Enforce and at times develop policies
- Be well founded in the legalities of healthcare
- Have finely tuned budget and financial management skills
- Manage union and nonunion employees
- And much more

Overwhelming? Yes, but doable because there are proven techniques that Nurse Managers use every day to tackle what seem to be insurmountable problems. *Nurse Management Demystified* describes those techniques and explains how to apply them in real-life clinical situations.

You might be a little apprehensive about pursuing a Nurse Manager's position, especially if you're an experienced clinician who witnesses first hand the types of problems that Nurse Managers are expected to resolve. However, by the end of this book, you'll be able to step up to management responsibilities on a unit and begin to solve practically any problem that comes your way.

A LOOK INSIDE

Nurse management is an overwhelming challenge unless you educate yourself with the help of books such as this, which highlights strategies that have been successfully used by current and aspiring Nurse Managers. Each chapter follows a time-tested formula that first explains techniques in an easy-to-read style and then shows how you can use the techniques in the real-world hospital environment. You can then test your knowledge at the end of each chapter to be sure that you have mastered the content. There is little room for you to go adrift.

Chapter 1: The Evolving Healthcare Delivery System

The healthcare industry strives to meet the health care consumer's expectations and, in doing so, faces complexity, high cost, and challenges that are not obvious to the general public. At the forefront of these efforts to provide high quality, cost effective health care are Nurse Managers, who direct a team of healthcare providers that deliver quality patient care within a healthcare facility that is faced with reducing expenses and increasing revenue in order to operate economically. This chapter explores the economic basis for rising healthcare costs.

Chapter 2: Nursing Leadership and Management

A Nurse Manager doesn't have the time nor all the necessary skills to care for patients and perform countless other tasks necessary to run a department. The Nurse Manager must organize and direct a staff to care for patients in a rapidly changing and culturally complex environment. This chapter teaches the leadership and management skills needed to become a Nurse Manager.

Chapter 3: Nursing Care Delivery Models and Staffing

A key ingredient in every patient care plan is a nursing care model, which defines a framework within which to care for patients. In this chapter, you'll learn how, as a Nurse Manager, to develop your own game plan around a nursing care model and how to use proven professional techniques to manage your staff members as they care for your patients.

Chapter 4: Delegation and Supervision

The Nurse Manager must develop and implement a strategy with which to delegate responsibilities to the staff members and then supervise them as they give patient care. Delegating responsibilities can be deceivingly simple. In this chapter, you'll learn how to delegate and how to supervise a staff under normal circumstances and in less than ideal situations.

Chapter 5: Effective Communication and Conflict Resolution

In this chapter, you'll discover how to avoid common problems by learning how to effectively communicate with your staff, colleagues, physicians, patients, and patients' families. You will also learn proven techniques for resolving conflicts that are bound to arise in your role as a Nurse Manager.

Chapter 6: Policy

Policies define practically everything you and your staff can do and how you can do it in the healthcare facility. Nurse Managers interpret and uphold policies and recommend new policies that address problems faced each day while caring for patients. This chapter explores policies in a healthcare facility and how policies are created, how to enforce policies, how to create work rules, and the politics of policies.

Chapter 7: Legal Issues

In a society that has become more and more litigious, the Nurse Manager must become increasingly aware of laws, rules, regulations, and policies that govern patient care. Failure to do so can result in legal action against the Nurse Manager, loss of license, fines, and incarceration in rare incidents. In this chapter, you'll explore the legal system and laws that govern healthcare. You'll also learn how to prepare for litigation and steps you need to take to reduce the likelihood that you will become a target of legal action.

Chapter 8: Healthcare Economics

You've heard the term economics before but perhaps not in the context of healthcare because most of us think of them separately. Healthcare economics is the system used to provide and pay for healthcare. In this chapter, you'll learn how the employer-based health insurance system has disrupted fundamental economic principles that keep prices under control. You'll also learn principles that you can use to manage your area of a healthcare facility through this radical change in healthcare economics.

Chapter 9: Budget Planning and Financial Management

The Nurse Manager runs both the healthcare side and the business side of a unit and must be as comfortable with finance and budgets as with patient assessments and treatments. In this chapter you'll learn about budgets, cash flows, trends, and other tools a Nurse Manager uses to manage the business side of a unit.

Chapter 10: Unions, Management, and Employee Relations

The nursing shortage, increasing demand for healthcare by aging baby boomers, and efforts by hospital administrators to bring skyrocketing healthcare cost back to earth are pressuring nurses to do the once unthinkable—unionize. This chapter introduces you to collective unions, collective bargaining, and how to manage a staff of union and nonunion nurses.

Chapter 11: Time Management

Your job as a Nurse Manager is to get things done correctly and on time with the tools and staff that you have at your disposal. You don't have the luxury of letting things slip or be done halfway because the lives of your patients depend on you completing your activities. In this chapter, you'll learn proven techniques that successful Nurse Managers use to manage their time both at work and at home.

Chapter 12: Nursing Informatics and Measurement

Information overload is eased by using informatics and measurement to filter out unimportant data, leaving only the information you and other healthcare providers need to care for patients. You'll learn about informatics and measurement in this chapter and see how by combining them with computer technology you can dramatically improve how you collect and use healthcare information.

Chapter 13: Risk Management

Some people are better risk takers than others because they know how to manage risk by reducing their exposure to situations where the risks outweigh the benefits. The Nurse Manager must become a smart risk taker in order to care for patients and manage staff. This chapter takes a close look at the risks involved with healthcare and how to manage those risks.

Chapter 14: Managing Scarce Resources

Money and registered nurses are just two of many scarce resources needed to care for patients. When there isn't enough of a resource to go around, the Nurse Manager must become creative and manage the scarce resource so that patients continue to receive a high level of care. In this chapter, you'll learn techniques to manage both scarce human and nonhuman resources.

Nursing management can be a stressful yet rewarding career choice. As a nurse, you have already learned to manage groups of patients and resources. As a Nurse Manager, you will perform those tasks on a larger scale with added responsibilities. This book is designed to give you the skills and resources to be successful as a Nurse Manager. Use it as a foundation on which to build and enhance your career, as well as your personal satisfaction with your ever-adverse, and always wonderful profession.

CHAPTER 1

The Evolving Healthcare Delivery System

Consumers who are insured tend to take the healthcare delivery system for granted and expect high quality treatment at no out-of-pocket expense. When ill, people want to visit the most convenient healthcare provider and wait less than an hour to receive a cure-all pill that restores them to health by the time they arrive home. And no one expects to receive a bill.

The healthcare industry strives to meet these expectations and, in doing so, faces complexity, high cost, and challenges that are not obvious to the general public. At the forefront of these efforts are Nurse Managers, who direct a team of healthcare providers to deliver quality patient care within a healthcare facility that is trying to reduce expenses and increase revenue in order to operate economically.

Nurse Managers are leaders who set goals for the delivery of high quality patient care, establish procedures for the safe, cost-effective delivery of that care, and oversee the complex management of a healthcare unit. They also operate the business and administrative areas of a healthcare facility. A Nurse Manager troubleshoots medical and administrative problems, and then devises a game plan and coaches a team of healthcare professionals to work toward a solution to these problems.

The challenge faced by Nurse Managers is enormous. The reward is fulfilling, though, because out of the chaos of today's healthcare system, a Nurse Manager finds a way to meet both patients' expectations and the expectations of administrators who run the healthcare facility: deliver quality healthcare economically regardless of staff shortages, higher expenses, stringent insurance coverage, and poor medical reimbursement.

This chapter explores the economic basis for healthcare as provided in the United States, and the role of Nurse Manager as well as the challenges they face today. You'll learn techniques on how to meet these challenges throughout the remaining chapters of this book.

Rules and Regulations in the Healthcare System

The healthcare industry is one of the most regulated industries in the United States. Its revenue is strongly influenced by guidelines established by Medicare, Medicaid, and medical insurers. In addition, there are regulations that govern the operation of healthcare facilities.

Healthcare facilities must be accredited by an accrediting organization that establishes standards for providing healthcare. The Joint Commission on Accreditation of Healthcare Organizations (JCAHO) is the most prominent accrediting organization. Every three years, its staff grades a healthcare facility based on the facility's compliance with the organization's standards. The grade is available to the public. Only institutes who receive a passing grade are accredited, enabling the facility to receive reimbursement from the government and private insurers.

It is within this highly regulated environment that a Nurse Manager participates in the setting of policies and procedures to provide safe and efficient patient care. A *policy* is a rule that governs the operation of a unit, department, or division of a healthcare facility. For example, a healthcare facility may have a policy that permits the deputy director of nursing or a supervisor to reassign nurses during their shift to another unit based on patients' needs.

A *procedure* is a guideline for performing a task, such as a nursing procedure. For example, there is a procedure for changing I.V. tubes, which is based on the healthcare facility's experience. *Protocols* are established procedures to use in specific situations.

Policies and procedures enacted by a healthcare facility typically stem from standards or regulations of the accreditation organization and of federal, state, and local governments. Federal regulations generally focus on public health, welfare, Medicare and Medicaid, social services, and bioterrorism, and include rules that cover the administrative aspects of a healthcare facility, such as accounting, labor, and wages.

State governments focus on protecting the public by licensing healthcare professionals and healthcare facilities. States also determine the number and types of healthcare facilities.

Publicly Operated vs. Privately Operated Healthcare Facilities

There are generally two categories of healthcare facilities: privately operated and publicly operated.

Privately operated healthcare facilities can be for-profit or not-for-profit institutions. A *for-profit healthcare facility* is owned by an individual, partnership, or corporation for the purpose of providing healthcare in return for a profit. Stockholders' profits are an important consideration in a for-profit institution. Indigent care can suffer with this type of healthcare facility, as an indigent patient will be transfered to a city or state supported facility as soon as it is safely possible.

The owner of a for-profit healthcare facility invests money to acquire the facility, equipment, staff, and the necessary licenses to operate the facility. The owner anticipates that, over time, the facility will take in more money than it pays out, resulting in a return on the owner's investment.

A *not-for-profit healthcare facility* is owned by a not-for-profit organization whose sole purpose is to provide healthcare to the community. Funding for the initial investment comes from grants, loans, and fundraising. Operating expenses are covered by insurance reimbursements and donations. A not-for-profit facility is expected to generate a surplus of funds, and reinvest those funds into providing additional services to the community. Many are owned by towns and counties, while others are still owned and operated by the communities they serve.

A *publicly operated healthcare facility*, such as a Department of Veterans Affairs (VA) hospital, is typically operated by a government agency and operates as a not-for-profit institution. Although public healthcare facilities strive to cover expenses through revenue generated from insurance reimbursements, many have shortfalls—expenses are higher than revenue—that are filled by taxes.

Organizational Structure of a Healthcare Facility

A board of trustees is responsible for the total operation of the healthcare facility. Trustees are elected or appointed by the owner of the facility. In a government-operated facility, trustees are typically appointed by the governing body, such as the mayor and local council, or the governor. Trustees of privately operated facilities are elected (corporation/partnership) or appointed (individual) by the owners.

The general duties of the board of trustees are to:

- Establish a mission, vision, and philosophy
- Set long-range goals
- Establish long-range capital-improvement plans and related financing
- Approve and monitor the operating budget
- Appoint and oversee a chief executive officer (CEO)
- Appoint executives to work with the CEO to achieve the mission's goal
- Approve admitting privileges for physicians
- Oversee the operation of the entire facility

The CEO is the president of the facility and is responsible for enacting policies created by the board of trustees. The CEO hires a team of executives consisting of the chief operating officer (COO) and senior management, who are sometimes referred to as senior vice presidents of the facility. The COO handles day-to-day operation of the facility to ensure that the staff achieves goals established by the board of trustees. Senior management personnel report to the COO.

Each senior manager is responsible for the operation of one or more departments of the facility. A senior manager is expected to establish policies within the guidelines provided by the COO, CEO, and the board of trustees, to operate the department efficiently and to achieve the goals set by the board of trustees. Senior managers typically look five years ahead to determine the healthcare demands and the challenges to delivering healthcare that lie ahead for their departments. They then develop a plan to meet those demands and challenges.

The following are the departments that are common to most healthcare facilities:

- Patient care (nursing), respiratory therapy, physical therapy, etc.
- Facilities (engineering)
- Radiology
- Laundry

- Transportation
- Admitting
- Medical board office
- Social services
- Housekeeping
- Pharmacy

Department directors are common in large healthcare facilities and are responsible for a subgroup of departments within the area of a senior manager. Department directors plan for events that are expected to happen two to five years out for each department. A department director also establishes departmental budgets, approves spending, and approves staffing. In some larger healthcare facilities, responsibilities of a department director are delegated to one or more assistant department directors.

A department manager, also known as a unit manager, is the front-line manager responsible for the day-to-day operation of a department (unit). A department manager handles work schedules, daily problems, and short-term planning (less than two years). Larger healthcare facilities employ assistant department managers, who share some of the department manager's responsibilities, such as supervising a shift.

The staff follows policies and procedures that are established by the management team (board of trustees, CEO, COO, vice presidents, department directors, and department managers) to provide economical patient care at a level that meets the board of director's mission statement.

A Nurse Manager typically fulfills the role of a department/unit manager and assumes responsibility for the day-to-day operations of a department/unit 24 hours a day, seven days a week. For example, a Nurse Manager for a 12-bed pediatric department might supervise an evening and night charge nurse, plus 15 registered nurses, 10 nonprofessional staff, and an annual budget of $1.5 million.

In comparison, a Nurse Manager responsible for a medical cardiac care unit of 36 beds might have three assistant department/unit managers, or for each shift, days, evenings, and nights, a nurse practitioner, 24 registered nurses, 11 nonprofessional staff, and an annual budget of nearly $3 million.

Challenges Facing Hospitals

Hospitals and other healthcare facilities are being crunched from two sides. Baby boomers have an ever-increasing demand for more healthcare. Medical insurers demand lower healthcare costs. The healthcare facility is in the middle trying to meet both demands, and the Nurse Manager is at the forefront of this struggle.

REIMBURSEMENT

The U.S. federal government uses a retrospective payment scheme called Diagnosis Related Groups (DRGs) to identify care given to Medicare and Medicaid patients. *DRG* is a patient classification system that relates the cost of treating similar types of patients within a diagnosis to ultimately set a common reimbursement. The hospital is reimbursed a specific dollar amount for each DRG by the federal government for Medicare, and by a combination of state and federal governments for Medicaid. However, the fee for each DRG has been declining in recent years as both federal and state governments face budget crises.

Managed Care Organizations (MCOs) and Health Maintenance Organizations (HMOs) have been seen as a viable way to manage healthcare costs for medical insurers. Patients who join these organizations are fully covered for healthcare as long as they use healthcare facilities and healthcare providers who are members of the organization. To remain competitive, hospitals enter into contracts with these organizations, making the hospitals available to patients who join an MCO or HMO. However, hospitals must agree to a negotiated, and usually lower, reimbursement.

The rise in healthcare costs has forced an increasing number of employers to reduce or drop medical coverage for their employees. As a result, a growing number of people lack medical insurance. Many of these people don't go to a physician when they become sick. Instead, they take home remedies or simply hope the illness resolves itself. When it doesn't, they frequently end up in the hospital. Hospitals cannot turn away patients. Even private hospitals must stabilize a patient before the patient can be transferred to a public hospital for treatment.

Hospitals are seeing their costs increase and their revenues drop because of low reimbursements and noncovered patients. This situation forces Nurse Managers to devise innovative ways to achieve efficiencies in operating the department without sacrificing the high-quality healthcare provided to patients.

JCAHO STANDARDS

JCAHO establishes operating standards for hospitals. JCAHO's goal is to require all hospitals to have policies and procedures in place that ensure patient, staff, and visitors' health and safety while in the hospital. This is accomplished by requiring health care organizations to implement myriad policies on everything from infection control and staffing, to medical staff education.

While few can disagree with JCAHO's goals, these standards can have a dramatic impact on the hospital's operation, because many of them increase the operating cost of the hospital. This increase in cost is not offset by an increase in reimbursement.

In addition, the Nurse Manager and the staff must keep abreast of changes to the JCAHO standards, and implement those standards within JCAHO-specified deadlines in order for the hospital to remain in compliance.

STAFF SHORTAGE

Many hospitals are experiencing a shortage of nurses, pharmacists, and other healthcare workers. For a long time, those professions were less attractive, with lower salaries and fewer benefits, than careers outside the healthcare industry. The tide is beginning to change. There is an increasing number of applicants for healthcare training programs, but many are turned away because there aren't enough qualified instructors.

The shortage of nurses places a strain on Nurse Managers, who must cover the department 24 hours a day, 7 days a week, with a limited number of nurses. At times the shortage can result in nurses being moved from one department to another during a shift in order to meet heavy demand for care.

MEDICAL ERRORS

According to the Institute of Medicine Report, titled "To Err Is Human" (11-01-99), it is estimated that 90,000 patients die every year because of medical errors that occur in hospitals. Many of these deaths are preventable, if the staff follows policies and guidelines that are adopted by the hospital and the hospital's board of trustees.

Medical errors may occur when the nurse is under time pressure to perform procedures on several patients nearly simultaneously. To provide treatment, the nurse makes certain assumptions that are frequently valid, but not always. For example, the nurse may administer medication to a patient without taking the time to learn about the medication. The assumption is that the patient received the same medication from the previous shift, and therefore there was probably nothing wrong with the medication order. A common time for medication administration is 9:00 A.M. A nurse may have several medications due at the same time. If these medications are not given within the acceptable window of time, the nurse can be charged with a medication error. A medication error could also occur if not all procedures are followed when giving medications.

It is the responsibility of the Nurse Manager to create a work environment that stresses the importance of following policies and procedures. Furthermore, the Nurse Manager needs to alleviate obstacles that prevent the staff from fully implementing these policies and procedures.

RISK MANAGEMENT AND LITIGATION

Hospitals are fertile grounds for litigation. Patients are placed in the care of medical staff who often must perform invasive procedures that expose the patients to risk of injury. If patients believe that they have received care that is below acceptable standards, or even harmful, they may well bring suit against a healthcare facility. Some patients look to the courts for monetary relief for pain and suffering that the patient allegedly received at the hands of hospital staff.

The Nurse Manager cannot eliminate the risk of litigation. However, they can manage the risk that a patient will be injured while being treated in the hospital, by stressing patient safety initiatives and utilizing patient safety policies.

Nurse Managers and staff members must identify risks and report untoward events to the hospital's risk management department. Risk management staff participates in correcting potentially hazardous situations, and suggests action to either eliminate or reduce the hazard. For example, a common risk is that a patient will fall while in the hospital. The Nurse Manager can reduce this risk by carefully investigating incidents of patient falls to determine why the patient fell, and then instituting policies and procedures to avoid conditions that could lead to further incidents. The investigation would note the time of day that falls occur, the age and condition of the patients, medication the patients were given prior to the fall, which footwear the patients were wearing, and other factors that could have led to the falls. Nursing staff can also help by identifying those patients who might be at risk, by developing or using some sort of fall risk scale.

The Nurse Manager must also make sure that the staff files an incident report whenever an untoward event occurs during the shift, even if patients are unaware that the event occurred. Incident reports are used to identify potentially hazardous conditions that need further investigation. In addition, an incident report alerts the hospital's legal team to a situation that might lead to litigation. Facts related to the incident are gathered and form the basis for the hospital's defense.

PERFORMING PROFITABLE PROCEDURES

A hospital's main goal is to provide quality care to patients. To do this, it must bring in revenue and keep down costs to generate a surplus of cash. (The surplus is referred to as *profit* in a for-profit hospital.) Typically, hospital administrators, along with physicians, identify procedures and services that are both profitable and in demand, such as same-day-surgery facilities where a patient arrives early in the morning for surgery and goes home in the evening. Hospitals try to eliminate unprofitable services or maintain them with profits from other services. For example, a hospital may deliver too few babies to earn a profit on Labor and Delivery, but maintain the service as it is needed in the community. The loss will be made up by a profitable cardiac surgery program, that is both cost effective and a community need.

Hospital administrators, in conjunction with physicians, identify a procedure and the medication and other supplies that are necessary to perform the procedure, based on expense and effectiveness for the patient, and the profitability for the hospital.

It is the Nurse Manager's responsibility to make sure that, from a nursing perspective, the procedure is performed according to guidelines established by the hospital, and that medication and other supplies used in the procedure are not wasted.

The result of careful planning by the hospital's administration and the carrying out of those plans by the Nurse Manager is that the expected cash surplus from the procedure will be realized by the hospital.

TECHNOLOGY COSTS

Computers are playing an ever-increasing role in healthcare. Computers are used to monitor patients, perform laboratory studies, and manage patients' medical records. An increasing number of physicians are using computers to write orders for their patients. Likewise, nurses use computers to chart the patient's progress.

Hospitals incur an enormous expense to outfit departments, patients' rooms, and administrative offices with the latest computer technology that enables information to be shared among staff throughout the hospital, and with physicians and other healthcare professionals located outside the facility.

In addition to computer technology, hospitals must acquire and maintain a long list of equipment such as lasers, operating room equipment, and specialty equipment used at the bedside. The typical hospital is finding itself in a technology crunch. Patients expect that the hospital will use the latest technology to provide them with healthcare and treatment that leads to the best possible outcome. If the hospital doesn't, then patients will likely choose a different hospital. However, technology is expensive and constantly changing.

COST OF MEDICATION AND SUPPLIES

Hospitals are reimbursed by procedure or by the length of time a patient stays in the hospital, not according to what it costs the hospital for the stay. This means that the hospital cannot pass along higher costs to third-party payers if the cost of the entire stay increases.

Hospital administrators carefully negotiate terms with third-party payers so that reimbursement covers the cost of treating the patient. The Nurse Manager must make sure that there isn't any waste when treating the patient. This is difficult at times, because typically the staff isn't aware of the cost of medications, supplies, or equipment. The Nurse Manager must constantly weigh the cost of patient care with the necessity of providing optimal care. Nurse Managers must help the staff understand the financial side of healthcare.

PERSONNEL COST

The cost of hospital personnel has dramatically increased in recent years, partly because of staff shortages, and partly because of union demands for better pay and benefits, as well as work rules that limit work schedules. Staff shortages result in competition among hospitals to attract a limited number of qualified employees. Offering higher salaries and benefits is the way many hospitals attract new employees.

Staff shortages have also forced hospitals to create a variety of costly work schedules. Some hospitals require employees to work mandatory overtime on very short notice, and assign employees to different departments during a full or partial shift. Nurse-to-patient ratios increase, making it practically impossible for nurses to have any breaks during the shift. Many institutions use the services of nurse agencies. Hospitals who hire these agency nurses are not required to provide the same benefits as they do for their own employees. They are hired for a specific time-frame. However, agency nurses usually charge more than what hospital employee nurses make.

Burdensome work schedules cause some staff to join a union, in order to negotiate better work rules, salary, and benefits.

COST OF INSURANCE

With an increasing number of patients suing hospitals and physicians, and with jury awards skyrocketing into million of dollars in some instances, insurers have increased premiums to a level that sometimes makes the delivery of healthcare unprofitable or impossible.

Hospitals are being squeezed on one side by insurance companies that are increasing premiums and reducing insurance payments to the hospital, and on the other by physicians who have admitting rights at the hospital demanding that the hospital pick up some or all of the physicians' malpractice insurance if the hospital wants the physicians to continue to admit patients.

DECREASING LENGTH OF STAY

A hospital is not like a very expensive hotel where the longer a guest stays, the more money the hotel makes. Healthcare insurers want patients to stay the shortest possible time in the hospital, and will reimburse a set fee regardless of the patient's length of stay. On one hand, the hospital receives a more profitable reimbursement for a one-day stay, but on the other, the hospitals need a greater number of patients to generate the revenue required to support the operation of the hospital.

In essence, this means that hospitals need to find more patients. To do this, some hospitals have merged with other healthcare facilities, and others have expanded into new services, such as open heart surgery. Marketing has become very important in attracting patients to one hospital rather than another. Hospitals will advertise in the local media and stress their particular strengths. For example, open heart surgery without the use of blood transfusions, or wonderful maternity care based on the high number of deliveries. However, hospitals, unlike other businesses, must apply for a Certificate of Need to the state government to expand their services. The government requires this to ensure that healthcare resources are focused on the healthcare needs of the population, and not directed toward services where demand is minimal.

The move toward reducing the length of hospital stays can have a negative impact on the quality of patient care. For example, early discharge deprives a patient of the optimum number of rehab sessions. This may also place a stressful burden on the family, which is now required to provide complex care to the patient at home. The Nurse Manager and the staff have an enormous impact on the length of time a patient remains in the hospital because, to a large degree, they carry out orders in a timely fashion and encourage patients to comply with activities that enhance recovery, such as deep-breathing exercises, coughing, and ambulation. For example, the physician may state that the patient can become ambulatory with assistance eight hours after returning from surgery. However, nursing staff availability actually determines when the patient becomes ambulatory. If there is a shortage of staff, or the staff is busy with other patients, then there may be a delay getting the patient out of bed. The result is possibly a longer stay in the hospital.

The Nurse Manager can assure that the patient's stay is as short as possible by using a *care map*. A care map is a document that projects the progress of a patient. Using evidence-based research, the patient's daily progress can be forecast and then compared with their actual progress. At any point during the patient's care, the Nurse Manager will know if the patient is on schedule or behind schedule, and then can make any necessary adjustment to bring the patient back on track.

DEMOGRAPHIC CHANGES

The *demographics* of the area serviced by a hospital are simply the hospital's customer base. Many hospitals are experiencing a population shift that is altering their demographics. For example, as the baby boomers retire, they move from private medical insurance coverage to Medicare, which may have lower reimbursement to hospitals than private medical insurers. The result is lower revenue for hospitals. In the 1990s, home health agencies increased in number. This was in part to help with those patients who were leaving the hospital earlier and still needed nursing care. Many hospitals opened their own agencies to provide another source of income for the hospital. Before long, legislation was passed to limit the number of home health visits that Medicare would reimburse. As a result, many home health agencies were forced to shut their doors, unable to meet the expenses of operating a healthcare business.

Some hospitals are witnessing an increase in uninsured patients due to the high cost of health insurance, and an increase of low wage earners whose employers traditionally don't offer medical coverage.

Challenges Facing Long-Term Care Facilities

Long-term care facilities are facing some of the same challenges that hospitals face. Few patients are able to afford to pay for long-term care. Therefore, nearly all revenue comes from Medicaid reimbursements. Many times, patients are forced to liquidate their assets in order to qualify for Medicaid.

Staffing is also a challenge for long-term care facilities. The facility must have at least one registered nurse on duty 24 hours a day, 7 days a week, as well as certified nursing assistants. Typically, the facility must train its own nursing assistants. The facility might also need to employ physical therapists, occupational therapists, recreational therapists, auditory therapists, and speech therapists. The cost of these employees is not always fully covered by Medicaid reimbursement.

Long-term care facilities must also meet tough federal and state regulatory standards, which increase the cost of operating the facility.

Challenges Facing Residential Care Facilities

A *residential care facility* provides shelter and assistance to patients who are afflicted with debilitating conditions, such as mental retardation or cerebral palsy, but don't require constant medical care. These patients are very difficult to care for in terms of meeting their physical and psychological needs. Nearly all revenue for the facility comes from reimbursement from Medicaid.

The facility isn't required to employ a registered nurse or a certified nurse's aide. The staff is usually low paid and low skilled, and therefore it is difficult to attract quality workers. One of the most difficult challenges residential care facility managers face is finding and maintaining a location for the facility. Everyone agrees that there is a need for residential care facilities, but no one wants a residential care facility in their neighborhood.

Challenges Facing Ambulatory Care Centers

An *ambulatory care center* is a general term that describes various kinds of facilities where patients are treated and leave the same day. These include physician's offices, clinics, birthing centers, same-day surgical centers, imaging centers, and endoscopy centers.

Ambulatory care centers are cost effective because they have a relatively low overhead and a high volume of patients. The volume of patients is particularly critical for centers that use expensive equipment to pay for diagnostic tests such as computerized axial tomography (CAT) scans. The cost per test is lower if a large number of tests are performed regularly.

However, ambulatory care centers receive most of their revenue from medical insurers, Medicare, Medicaid, and HMOs. All of these third-party payers reimburse at a rate lower than the billing rate of the center. Some states are requiring the ambulatory care centers located in their state to be licensed as a hospital, which requires standards that increase the costs, such as being open 24 hours per day, and having an RN on the premises 24 hours a day, 7 days a week.

Challenges Facing Home Healthcare Agencies

Home healthcare agencies provide in-home healthcare to patients who are no longer eligible for care in a hospital or other facility. These include patients who were recently discharged from a facility, or patients with long-term medical needs that can be provided by a nurse or other healthcare professional at home.

Home healthcare is not fully covered by Medicare or Medicaid, thus requiring patients to personally pay for care, or use private medical insurance for coverage. This limits the customer base for home healthcare agencies.

Staffing is another challenge, because healthcare providers tend to desire positions in healthcare facilities, rather than having to travel to five or more patients' homes every day, where they have little if any supporting staff. This situation is further complicated if the patient is on a respirator or requires intravenous therapy.

Challenges Facing Managed Care Organizations

A *managed care organization* is responsible for managing the healthcare of its clients, which are typically major employers such as IBM or a state agency responsible for Medicare. The recipients of the care are the employees of those clients, or patients whose medical care is reimbursed by the client.

There are different kinds of managed healthcare organizations, the most prevalent of which is the *health maintenance organization* (HMO). An HMO provides healthcare using one or multiple arrangements with healthcare providers. These arrangements are based on various models:

- **Staff model**   The HMO employs salaried physicians to treat patients. Physicians are offered various incentives based on their performance and productivity.

- **Group model**   Groups of physicians, each group representing a specialty, contract with the HMO to provide healthcare for a negotiated fee. The physicians are employed by the group practice, not by the HMO.

- **Independent practice association (IPA) model**   An HMO contracts with an IPA (a "middle man," of sorts) to provide care to members for a negotiated fee. The IPA contracts with physicians, who may be individuals or members of a group practice, and they are compensated on a fee-for-service basis.

- **Network model**   Physicians who have a group practice contract with an HMO to provide healthcare for a fixed monthly fee per patient. This is referred to as a *capitated* fee.

Managed care organizations have an advantage over hospitals and other healthcare facilities, because they receive a guaranteed monthly payment from their clients, and their costs to provide healthcare are fixed by contracts that they have with healthcare providers.

However, managed care organizations are faced with a serious challenge: patients prefer to use their personal physician. Their personal physician may not have a contract with the managed healthcare organization. This means the patient may complain to his or her employer, which in turn may seek the services of a different managed healthcare organization. Another issue is patient volume. Sometimes the patient volume is so high that it is either difficult for a patient to see a provider, or the provider must limit new patient referrals.

Challenges Facing Preferred Provider Organizations

A *preferred provider organization* (PPO) is similar to an HMO in that the PPO contracts with independent healthcare providers to provide care at a discount rate. These providers are considered "in-network" providers. The patient has a choice of going to a healthcare provider who has a contract with the PPO, or going to any healthcare provider. If the patient goes to a contracted healthcare provider, the patient does not pay anything. If the patient goes to a healthcare provider that doesn't have a contract with the PPO, the patient pays an expensive co-payment.

One version of a PPO is called the *point-of-service plan* (POS). The POS has a very large network of healthcare providers who have contracted with the POS. A POS plan commonly has an HMO component along with indemnity-type coverage. That is, the patient pays the physician directly, and then sends the bill to the plan for reimbursement.

Work Rules and Unions

The board of trustees of a healthcare facility adopts rules regulating the personal conduct of employees. These rules are referred to as *work rules*. Some work rules apply to all employees, such as rules regarding dress code, health and safety, falsification of documents, and work hours. Other work rules apply only to employees who are members of a union that negotiated an employment contract with the healthcare facility.

Nurses and other healthcare professionals can be burdened by work rules imposed upon them by a healthcare facility. Individually, they are practically powerless to change a work rule, because they can be terminated without cause at any time by the healthcare facility.

In 1946, the American Nurses Association (ANA) moved to strengthen the power of nurses by adopting an economic security program that encouraged local associations to act as bargaining agents for nurses. The ANA became the first union to represent nurses.

A union is an organization formed by employees to negotiate work rules, wages, and benefits on behalf of member of the union. Thirty percent of the employees at a specific organization must request the National Labor Relations Board (NLRB) to conduct union elections as the initial step of forming the union. The NLRB is the arm of the federal government that oversees labor unions.

If fewer than 30 percent of the employees request an election, then an election cannot be held and each employee negotiates on his or her own behalf. Once an election takes place and the majority of employees vote to form a union, then the union becomes the official bargaining agent for those employees who join the union. The healthcare facility is obligated to recognize the union as the legal representative of its union members. Employees can form their own union, but typically they elect to have an existing union such as the ANA represent them.

Unions negotiate the following on behalf of their members:

- Rates of pay
- Wages
- Hours of work
- Conditions of work
- Grievance procedures

Unions may negotiate the following as well:

- Seniority
- Fringe benefits
- Management rights
- Floating
- Layoffs
- Professional issues

The reasons that typically lead nurses to unionize are the perception of being paid low wages, inability to communicate with management, authoritarian behavior of supervisors, and perceived abuse of existing work rules, such as mandatory overtime, floating patient care assignments, and no procedure for pursuing grievances.

Healthcare facilities can avoid unionization by having open communications with the nurses and fair work rules that don't overly burden the staff. There must be a fair and objective procedure in place to hear and resolve staff grievances. Failure to take these steps might result in disgruntled and discontented staff that is left with two choices: resign or unionize.

Once a union is formed, both the union and the healthcare facility are expected to bargain in good faith toward a contract. The contract defines the role of nurses and other union members covered by the contract, and defines the obligations of the healthcare facility. Many times, both sides come to terms within a reasonable period of time. However, occasionally an agreement cannot be reached and union members vote to go on strike. A strike is a work stoppage where union workers who are without a contract stop working and won't return to work until a new contract is signed by both the union and the healthcare facility.

A strike can be a good thing for a union, because it places pressure on the healthcare facility to come to terms. Sometimes a strike isn't such a good idea, because without a signed contract, the healthcare facility has the right to hire replacement workers, who may continue employment after the strike ends. When this happens, union employees can return to work based on seniority only when positions open up.

There are advantages and disadvantages to belonging to a union. Members of a union are treated as a unit—not as individuals. A superior nurse is treated the same as an inferior nurse. They receive the same pay and have the same opportunities for advancement.

Leaders who are elected by a majority of members represent a union. A member has to go along with the majority. So if the majority wants to strike, even those members who don't want to strike must do so. Union members must pay whatever dues the leadership requires, which at times can be a significant amount.

The Nurse Manager must be familiar with the details of the union contract, because the contract defines the work rules for the department. The Nurse Manager is not authorized to modify work rules or create new work rules. Nurses and other unionized staff can't modify work rules either.

Suppose the contract specifies that the healthcare facility must give two days' notice when requesting overtime, and nurses must volunteer for such overtime. A Nurse Manager may ask a nurse on the evening shift to also work the night shift. The nurse may decline the request because that is not two days of notice. If the nurse voluntarily agrees to work the overtime shift, that is not a violation of the contract.

Summary

Healthcare facilities are in the center of a struggle between patients who demand high-quality healthcare and third-party payers who seek to cut medical costs. Nurse Managers are at the forefront of this struggle and are responsible for keeping down medical expenses, while providing quality healthcare to patients.

Healthcare facilities must be accredited by an accrediting organization that establishes standards for providing healthcare. JCAHO is the most prominent accrediting organization. Every three years, its staff grades a healthcare facility based on the facility's compliance with the organization's standards. The grade is available to the public. Only institutions that receive a passing grade are accredited, enabling the facility to operate.

The two general categories of healthcare facilities are privately operated and publicly operated. Privately operated healthcare facilities can be for-profit or not-for-profit. A not-for-profit healthcare facility is owned by a not-for-profit organization whose sole purpose is to provide healthcare to the community and to generate a surplus of funds for the continuation of that purpose.

The board of trustees of a healthcare facility adopts work rules to regulate the personal conduct of employees. Some work rules apply to all employees, and others to members of a union that negotiated an employment contract with the healthcare facility.

Now that you have an overview of the environment within which a Nurse Manager must work, Chapter 2 takes a look at leadership and management techniques that the Nurse Manager can use to achieve the goals of the healthcare facility.

Quiz

1. The acronym JCAHO stands for:

   (a) Joint Commission on Accessible Healthcare Organizations

   (b) Joint Commission on Accreditation of Healthcare Organizations

   (c) Joint Commission on Accreditation of Health and Operations

   (d) None of the above

2. A guideline for performing a task is called a procedure.

   (a) True

   (b) False

3. The duties of the board of trustees are to:

   (a) Establish a mission philosophy

   (b) Set long-range goals

   (c) Approve and monitor the operating budget

   (d) All of the above

4. Medical errors are most often caused by:

   (a) A single person's mistake

   (b) Poor handwriting

   (c) Care-giver exhaustion

   (d) Systems that break down at various points

5. Revenue shortfalls in a government run healthcare facility are filled by contributions from taxpayers.

   (a) True

   (b) False

6. Department directors in a large healthcare facility are responsible for:

   (a) Patient assessments

   (b) Case management

(c) Administering medication

(d) None of the above

7. A not-for-profit healthcare institution must bring in less revenue than expenses.

(a) True

(b) False

8. What challenge do hospitals confront?

(a) Higher staffing costs

(b) Shortage of staff

(c) Low reimbursement

(d) All of the above

9. A Nurse Manager can change work rules that were agreed to in a union contract.

(a) True

(b) False

10. A hospital has the right to replace striking workers with new, nonunion employees.

(a) True

(b) False

CHAPTER 2

Nursing Leadership and Management

All the results of good nursing…may be spoiled or utterly negatived by one defect, viz.: in petty management, or in other words, by not knowing how to manage that what you do when you are, shall be done when you are not there.

—Florence Nightingale, *Notes on Nursing*, 1860

Want something done right? Do it yourself. This is not possible if you are a Nurse Manager because you don't have the time, and perhaps the skills required, to care for patients and perform countless other tasks necessary to run your department.

A Nurse Manager plans for efficient and economic healthcare for patients, and then organizes and directs the staff to carry out those plans in a rapidly changing and culturally complex environment. The Nurse Manager must know what has to be done, know when it must be done, and pick the right staff and tactfully coach them into doing the best possible job.

Some see the Nurse Manager as a ship's captain, the one who sets the course. The captain has knowledge of every job on the ship, but the crew provides the hands-on expertise necessary to safely and successfully take the ship to its destination. The Nurse Manager sets the tone for the unit. The captain provides leadership to the crew using intrinsic skills and learned behaviors, and manages problems that are encountered along the way to a destination that can be characterized as never static but always changing.

Management is not supervision, although some aspects of supervision are inherent in management. Much of a manager's time is spent planning for the success of a unit by setting both short- and long-term goals, by delegating appropriately, and by controlling the use of both material and human resources. In addition to a manager's personal knowledge and experience, outside factors affect the success of a manager. Among these factors is the organization itself; that is, the size of the organization, the level of support for first-line managers from senior management and human resources, the technology available to managers, the culture of the organization (whether it is hierarchical, flat, organized by product line, etc.), and the community in which the organization is located. The latter relates to the economic level of the population and the competition faced by the organization.

In this chapter you'll learn about the leadership and management skills that you need to become a Nurse Manager. First, consider how the role of Nurse Manager developed historically.

Nurse Management: A Look Back

Healthcare facilities grew in the United States as the population shifted from rural farms to urban cities, where people lived and worked in tight quarters, thereby fostering the spread of disease. The Pennsylvania Hospital of Philadelphia, which opened in 1752, was the first permanent general hospital in the United States.

The first hospitals were operated by physicians with donations provide by wealthy benefactors. As the demand for healthcare grew, so did hospitals, eventually growing into complex organizations that taxed the abilities of physicians to administer the hospital and care for their patients simultaneously.

Eventually a hospital administrator was brought on board to handle the administrative aspects of a hospital, freeing the medical staff to care for patients. The hospital administrator became responsible for nonmedical functions of the hospital—24 hours a day, seven days a week. The job was more than one person could handle, so caregivers were hired to manage daily activities of patient care. These caregivers were usually the lowest, poorest, and often sickest lower-class women, some of whom were prostitutes. Their management skills were poor. Hospitals were then known as a place to go to die—not to be cured.

By 1873, anesthesia and other new advances in medicine had made their way into hospitals. The perception of hospitals changed from a place to go to die to a place to get well. It was prior to and during this time that Florence Nightingale improved patient care by training caregivers in the skills necessary to bring the sick back to health. Caregivers were then called nurses.

The New York Training School for Nurses opened at Bellevue Hospital in New York City in 1873 and was the first school in the United States to train nurses. In 1888, the Mills Training School opened at Bellevue, the first school for the education of male nurses.

By the turn of the 20th century, hospital administrators ran the administrative aspects of hospitals, and nurses took care of patients around the clock. Hospitals created the position of *matron* to supervise nurses. A matron was an unmarried, middle-class woman who worked 12-hour days, six days a week, and lived in the hospital. The matron was the first Nurse Manager, and this was one of the few jobs during that period that a woman could hold to support herself.

After completing the two-year hospital-based nurse training, nurses received a diploma and went to work caring for patients. In the early 1940s, colleges began to offer high school students the opportunity to become nurses in a four-year program by earning a bachelor's degree in nursing.

Today, nurses can earn an associate degree, bachelor's degree, master's degree, and a doctorate in nursing. Each healthcare facility sets its own requirements to fill the position of Nurse Manager. The minimum requirement is to be a registered nurse. Some healthcare facilities require a Nurse Manager to hold a bachelor's degree or a master's degree in nursing.

Nurse Manager Roles and Management Styles

In general terms, a manager is a person who gets things done; however, this definition oversimplifies the role of a manager, because a manager must do many things to achieve the goal of getting things done. The role of the professional Nurse Manager is being reinvented at a rapid pace due to the addition of the "many things" that must be accomplished. Today's healthcare environment requires the Nurse Manager to spend more time working across organizational boundaries to interact with peers and to negotiate partnerships within and beyond their own department. Negotiating skills become essential.

A manager participates in goal setting, or is given a goal and general guidelines to follow by an upper-level manager. The manager plans to reach this goal by performing a number of tasks. However, it is up to the manager to identify each task, determine how long each task takes to complete, determine the sequence in which tasks are performed, and determine resources (persons or things) that are needed to perform these tasks. Knowing all this, the Nurse Manager will then delegate the task to the appropriate staff member.

The following example relates to a staff nurse. Suppose that a nurse is given the goal to rehydrate a dehydrated patient by infusing I.V. NS. The nurse determines there are two tasks necessary to reach this goal: prepare the I.V. NS, and infuse the I.V.

The first task, preparing the I.V., begins by reading the physician's order, and ends when the I.V. is ready to be infused at the bedside. The result is a prepared I.V. setup. The duration of the first task is 15 minutes. Resources required for this task are the I.V. medication, I.V. tube, I.V. stand, heparin lock, needle, gauge, and tape. An electronic delivery device, such as an IVAC, or manual device, such as a Dial-a-Flow, is also necessary. Many times IVACs are not immediately available; time is lost just locating one. The nurse may have to manually calculate the drops per minute until one can be found. Also required is a nurse to prepare the I.V.

The second task is to infuse the I.V. This task begins when the NS is infused into the patient, and ends when the full dose has been infused. The result is that the person is hydrated. The duration is eight hours. Resources that are required for this task are a room, a bed, a prepared I.V. setup, the patient, and a nurse to administer the I.V.

It may sound strange to include the patient as a resource, but it isn't, because without the patient you cannot infuse the I.V. This seems obvious. Yet in reality a nurse may assume that the patient is available for the I.V. when, in fact, the patient may be scheduled for a test at that time. The resource (patient) is unavailable and so the second task cannot be completed.

The second task is dependent on the completion of the first task—you can't infuse the patient until the I.V. setup is prepared. If the first task is delayed, so is the second task. This example depicts a management issue for a staff nurse, and each of you likely has encountered similar situations in your careers. (The preceding example is a staff nurse role/responsibility.)

Rehydrating a patient is a simple goal when compared to the goal a Nurse Manager is given in a healthcare facility. A Nurse Manager's responsibility is to operate a department, or groups of departments, within the general guidelines set forth by upper management and the trustees of the facility. One aspect of this responsibility is to monitor the budget. The assigned department(s) will be within the manager's scope of clinical expertise within the facility, and the challenge will be to do everything necessary to operate the department(s) within the budget guidelines—an enormous challenge that is filled with lots of complexity.

As an example of budget management, suppose that the monthly budget report for a department has an unfavorable variance of $3,000 dollars on the budget line for surgical supplies. The Nurse Manager budgeted $3,000 per month, as this is a surgical unit, but the expense line notes $6,000. Why $3,000 over budget? The Nurse Manager must investigate this variance and provide the explanation to the director of the division. Did the unit have twice the number of patient days as was budgeted? Were there one or more patients admitted who required many additional dressing changes? Did the price of surgical supplies go up this month? Did the hospital change to a more costly supplier? Was it pilferage? Did the staff simply waste surgical supplies? The Nurse Manager must find the underlying cause of this variance. Did the staff fail to document patient usage? Staff must be reminded to document charges when using supplies. Many times, in an emergency, it is the responsibility of one person to document all the supplies used.

This section discusses the many roles of a Nurse Manager, and then describes the three main styles of management that a Nurse Manager may adopt when fulfilling those roles.

NURSE MANAGER ROLES

Peter Drucker, who is a recognized master of managing an organization, described in his book, *The Frontiers of Management*, a manager as having four basic functions:

- Establishes objectives and goals for each area and communicates them to the persons who are responsible for attaining them.

- Organizes and analyzes the activities, decisions, and relations needed and divides them into manageable tasks.

- Motivates and communicates with the people responsible for various jobs through teamwork.

- Analyzes, appraises, and interprets performance, and communicates the meaning of measurement tools and their results to staff and superiors. Develops people, including self.

Drucker's list describes well the roles of a Nurse Manager, which are explained next.

Planner

Planning is one of the basic functions of management. Thoughtful planning will result in a blueprint, showing the action steps necessary for the achievement of future goals. The planner sets objectives to be accomplished and sets a timeline to note the progress of the plan. Nurse Managers generally work on short-term planning for the unit, but participate in strategic and long-term planning with upper management as well.

A good planner identifies what tasks must be performed; when to perform them; how each task is to be performed; who is performing the task; and the resources needed to complete the task.

Communicator

The staff must receive clear direction—what to do, when to do it, and how to do it—from the Nurse Manager. Failure to communicate effectively is a major reason why directions are not carried out as planned. A good communicator gives the staff detailed instructions to perform tasks that are necessary to reach the goal. Communication is an interpersonal process involving the sender and the receiver. The sender communicates something and it is not always good news. The receiver wants to hear good news, but should be prepared to receive all information, good or bad. Effective communication is concise and unambiguous.

Listener

A good Nurse Manager listens to the staff, patients, and everyone who interacts with the department. Listening provides the opportunity to receive valuable feedback that is used to avert some problems and resolve others. Active listening gives the Nurse Manager greater understanding of the issues being discussed. Preconceived ideas or prejudices should be put aside when listening to staff, patients, or families in order to be receptive to the truth. You cannot be a good communicator without being a good listener.

Decision Maker

A skilled decision maker evaluates all options before reaching a decision. A Nurse Manager analyzes factors leading to a decision, makes reasonable assumptions, and then makes a rational decision based on the facts of a situation—and devises contingency plans if some of those facts and assumptions prove false. This systematic approach allows the Nurse Manager to choose among alternatives, and to take prompt action on the decision. Communicating the decision process with staff can help prevent grumbling when implementing the decision.

Coordinator

Resources from all over the healthcare facility come together at the right time to provide healthcare to patients. These resources must be coordinated to assure that the right resource is available on time to perform each task that is necessary to treat patients and to keep the department operating smoothly. For example, clean linens may not be available until 10 A.M., which may hamper the staff when providing morning care. The Nurse Manager is the person responsible to discuss the situation with the Laundry Manager to devise a schedule that will meet the needs of the patients.

A successful coordinator will be well organized, with excellent time-management skills. The common goal should include excellent care for the patient provided in a cost-effective manner.

Delegator

The hardest thing to remember as a Nurse Manager is that you cannot do everything yourself. You must delegate tasks to the staff; provide guidelines for carrying out the tasks; and then verify that those tasks are completed. To delegate acknowledges to someone that you trust their ability to accomplish the task as well as, or better than, you could yourself. In appropriately delegating tasks to others, you free up time to focus on more complex aspects of running your department. As a practicing professional nurse, you learn rules of delegation with regard to other healthcare staff. They also apply here—delegate the right task, to the right person, with clear direction.

Supervisor

The American Nurses Association defines supervision as the active process of directing, guiding, and influencing the outcome of an individual's performance of an activity or a task. A good supervisor doesn't give orders to the staff. Instead, a good supervisor fosters a congenial work environment by correcting, praising, coaching, and training the staff to perform a superior job. Supervision goes on in the observation of the daily work of staff on the unit, and in a more formal setting in the yearly performance appraisal that is required by JCAHO.

Resources Manager

A Nurse Manager might feel there aren't enough staff, money, or other resources to provide cost-effective, quality patient care. This challenge can be solved by analyzing available resources and creatively deploying them in a manner that achieves the department's goals. Success might include utilization of a different care delivery model, or educating nonlicensed personnel to assume specific, nonprofessional duties.

Staffing Manager

JCAHO, along with other state and federal agencies, mandate that healthcare institutions provide adequate and qualified personnel. This means having the correct number of staff on each shift with the credentials to do the job. The Nurse

Manager prepares the schedule five or six weeks in advance and must be aware of policies regarding overtime, floating from one unit to another, and when to call on outside agencies for additional personnel.

The Nurse Manager must form a competent team and manage them to carry out the plan for reaching the goal of excellent care for patients and high productivity from staff. Self-scheduling is sometimes an effective way to adequately keep the unit staffed, and allows personnel to have a voice in deciding when they work.

Clinician

A Nurse Manager must have good clinical skills and judgment to handle any problem that cannot be handled by the staff. You must have some knowledge of the various diagnoses presented by the patients on your unit, as well as standard treatment protocols. The Nurse Manager provides clinical guidance to the staff, including when to call on the more expert help from clinical specialist nurses and physicians. As the staff members have the opportunity to see you as an expert clinician, they will also appreciate you as a manager and will listen as you guide them.

Evaluator

A Nurse Manager is inundated with information, some information that is important to making a decision, and other information that is a distraction. A Nurse Manager must evaluate each piece of information carefully before making a decision. Information will come from conversations with, and observations of, staff members, evaluation of equipment and procedures (such as the type of electric bed to be purchased, number and time of visiting hours, types of linens to be used, etc.), and, most importantly, an evaluation of oneself.

Leader

A leader inspires mentors and motivates the staff (as described in the upcoming section, "Leadership"). New York Mayor Rudy Giuliani showed leadership on and after September 11, 2001 by helping everyone deal with a horrific situation while his staff managed the disaster. The Nurse Manager must have leadership skills to effectively manage the staff during both times of quiet routine and times of chaotic change. A good leader does not have to be a good manager, but a good manager will be a good leader.

Troubleshooter

The Nurse Manager helps to troubleshoot problems facing their own and other departments. The Nurse Manager has the responsibility of providing all the resources (within the hospital's budgetary constraints) necessary for the patient care personnel to do their jobs effectively and efficiently. This responsibility extends to building relationships with managers in other departments to solve issues that impact on the patient care unit. For example, the Laboratory Manager might complain that specimens are arriving from the nursing department with handwritten labels rather than computer-generated labels. The Nurse Manager would be the person to discuss the issue with the Laboratory Manager. Perhaps the computers within the nursing department are unable to print labels, in which case the Laboratory Manager and Nurse Manager together would call upon the IT Manager to enable departmental computers to generate specimen labels.

Diplomat

The Nurse Manager must be a diplomat, because many times they lack the authority to directly resolve problems. For example, conflicts may arise between a physician and the staff. However, many physicians are not employees of the hospital, and therefore are not managed by an administrator. A physician is an independent practitioner who is a customer of the hospital. Therefore, the Nurse Manager must use diplomacy and conflict resolution skills to arbitrate and resolve such problems. (The exception is with in-house physicians, who are supervised by the hospital's medical director.)

A Nurse Manager must be mindful of the fact that physicians are the major source of patients, and thus the major source of revenue for the hospital. However, physicians also need admitting privileges to a hospital to care for patients, without which patients might choose another physician who has these privileges; in other words, admitting privileges are indirectly a major revenue source for the physician. Thus, it is in the best interest of both hospital administrators and physicians to build a cooperative working relationship to provide the best patient care—and maintain a good revenue flow. Many physicians will modify their actions if the Nurse Manager, either directly or through the medical director, diplomatically demonstrates how the physician's action hinders patient care.

Influencing a physician is easier if the Nurse Manager maintains open communication with physicians, which usually evolves over time into mutual professional respect.

PREPARING TO MANAGE

Reading about the Nurse Manager's role in the previous section probably has you wondering if you have the skills to become a Nurse Manager. You probably don't—at least not as a neophyte manager, and perhaps not all those skills to the same degree of proficiency. Few Nurse Managers are equally proficient in all areas of responsibility.

Nurse Managers hone their management skills through preparation—similar to the preparation that you received when you became a nurse. Throughout this book, you'll be presented with the knowledge that every Nurse Manager needs to manage successfully. Practice the principles you are learning, and soon they will be a part of your management repertoire.

It is important that as you progress through this book you acknowledge your weaknesses and then work to transform them into strengths. Be sure to allocate sufficient time to learn these managerial skills, and to practice them, which is the only way that you can accomplish all that is required to achieve success in the management of a department.

NURSE MANAGER STYLES

The manner in which a Nurse Manager supervises the staff of the department is referred to as a *management style*. This section discusses several common management styles that are adopted by Nurse Managers.

Authoritarian

The authoritarian style of management is one in which the Nurse Manager makes all the decisions and creates all policies with little if any input or feedback from the staff. The Nurse Manager dictates who will do the work and how the work will be done. The staff is given negative reinforcement when rules are not followed, and very little positive reinforcement when they are.

A new Nurse Manager who is insecure, and has little if any confidence in the staff to perform their duties, frequently adopts the authoritarian style of management because this provides the Nurse Manager with a sense of security and control over the staff.

The authoritarian style of management is successful in chaotic situations when tasks must be performed quickly, such as when a patient goes into cardiac arrest, because one person gives directions. The authoritarian management style is successful with employees who are being trained or who have yet to achieve a proficiency that enables them to work under general supervision.

The authoritarian management style has a serious drawback, though—no one likes to be micromanaged. Micromanagement occurs when the manager does not acknowledge the employee's capability to perform a task independent of supervision. Once a person is trained to perform a task, the person doesn't expect to be told how to perform the task, or in some cases when to perform it. For example, a nurse expects to be assigned patients by the Nurse Manager. However, a nurse often resents being told when to read a physician's orders and how to prepare medication.

Democratic

The democratic style of management encourages staff participation in planning and decision making, yet retains responsibility for making the final decision and devising the plan.

To adopt the democratic style of management, the Nurse Manager must have good communication skills and believe that the staff is always striving to do their best. The Nurse Manager provides general supervision and encourages the staff to take responsibility for their own work by providing positive feedback.

The democratic style of management works well with a trained staff. It is less successful for employees who have not achieved a proficiency that enables them to work under general supervision.

Laissez-faire

The laissez-faire style of management is one in which the Nurse Manager provides little guidance to the staff and does not stress the importance of following policies and procedures. The Nurse Manager seldom makes decisions, and accepts the status quo rather than seek innovation to improve patient care. The Nurse Manager who adopts a laissez-faire management style typically avoids personal contact with the staff, and communicates with them by using memos or e-mail. The staff is rarely provided with any positive feedback from the Nurse Manager.

The laissez-faire management style is frequently adopted by new Nurse Managers or Nurse Managers who are at the end of their careers. A new Nurse Manager often doesn't want to make changes, fearing that the change will disrupt the smooth working of the department. Or, a new Nurse Manager may simply be ill-prepared to manage. Nurses who have shown great proficiency in patient care often are promoted to a management position, but being a great nurse does not necessarily translate into being a great manager. New Nurse Managers need training and mentors to make the transition. An end-of-career Nurse Manager sometimes feels that there is no need or reason to change the department, because it will just be changed again when they retire.

The laissez-faire management style works well for a short time, if the staff is experienced in handling the day-to-day operation of the department. However, the Nurse Manager must become more active if the department evolves into a more complex organization.

Nursing Leadership

Well-known management guru Warren Bennis said, "Managers are people who do things right and leaders are people who do the right thing."

New York Mayor Rudy Giuliani's actions during the 9/11 crises clearly illustrate Bennis's point. City managers were doing things right by responding to the emergency situation. Mayor Giuliani did the right thing by showing New Yorkers, and the country, how to stay calm amid chaos and how to approach the aftermath of the disaster. Mayor Giuliani walked the streets of midtown Manhattan to show everyone that it was safe to return to their daily routine.

Leadership is the ability to guide and influence others without necessarily having the authority to direct their behavior. Mayor Giuliani led by doing the right thing. He walked the streets to show that they were safe and that it was time to get things back to normal. His actions were far more influential than if he had stood behind the podium in the command center saying that it was safe to go about daily activities.

This section describes the common leadership theories that may be applied to Nurse Managers.

TRAIT THEORY

Trait theory states that leaders have inborn traits, referred to as characteristics, that help them lead. These traits include intelligence, ambition, aggressiveness, self-confidence, orderly thinking, and flexibility.

BEHAVIORAL THEORY

Behavioral theory says that leaders are not born to lead, but learn leadership behaviors. These behaviors include the ability to create a structure within which schedules are set, procedures are written, and policies are decided, and to consider and respect the views of others.

CONTINGENCY THEORY

Contingency theory states that leadership behaviors are dictated by a situation. That is, a Nurse Manager demonstrates aggressive leadership in an emergency when it is critical that tasks be performed quickly and accurately. However, when working with a new nurse, the Nurse Manager adopts a nurturing behavior that gives time for the nurse to become acquainted with departmental procedures.

CHARISMATIC THEORY

Charismatic theory is a leadership style formulated by the emotional commitment of the manager to reach a goal. Mayor Giuliani displayed this leadership style when he showed his emotional commitment to New York City and to giving the country a sense of normalcy following the 9/11 attack. His enthusiasm spilled over to his staff, and everyone in the country, resulting in a return to order in the face of smoldering ruins and the retrieval of human remains.

TRANSACTIONAL THEORY

Transactional theory states that leaders work to maintain the status quo and give feedback to the staff only when an error occurs. A routine typically develops in a department that gives both the leader and the staff a sense of security and comfort. They interact with each other in a political sense where each expects to give and to receive a benefit, such as changing a work procedure for additional compensation.

TRANSFORMATIONAL THEORY

Transformational theory postulates that the status quo changes when the staff adopts the leader's vision and beliefs for reaching a goal. First the leader establishes trust and then demonstrates integrity. The staff is then empowered to work toward the goal. Top-level Nurse Managers frequently display transformational theory in action when they excite middle managers about their vision. This theory empowers middle managers to use their own initiative to develop policies and procedures to make that vision a reality.

SHARED LEADERSHIP THEORY

Shared leadership theory is a theory of co-leadership that states that the leader should form a partnership with the staff and empower the staff to manage themselves. This theory works well when teams are formed to care for a group of patients. Each team is self-directed and requires only oversight supervision rather than day-to-day supervision. *Oversight supervision* is where the Nurse Manager's involvement with day-to-day supervision of the group is minimal. This theory works with experienced and self-directed staff members.

Standards of Professional Practice and Professional Performance

The Standards of Professional Practice (six standards) and Standards of Professional Performance (nine standards) are a combined set of standards developed by the American Nurses Association (www.ana.org) to help Nurse Managers and the nursing staff provide quality healthcare to patients. The six Standards of Professional Practice are as follows:

- **Standard 1: Assessment**   Assesses effectiveness and efficiency of workflow and utilizes data collection in problem solutions. After a plan is executed, gather data and determine if the plan was effective and efficient. Use this information when formulating your next plan.

- **Standard 2: Problems/Diagnosis**   Identifies resources for decision making and promotes quality improvements in clinical areas. A Nurse Manager makes a decision after receiving information from various resources that are likely to have more knowledge than the Nurse Manager about a situation. Only then should a Nurse Manager reach a decision on a plan for promoting quality improvements in the clinical area.

- **Standard 3: Identification of Outcomes**   Encourages interdisciplinary teamwork in the improvement of patient care outcomes. It is important for the Nurse Manager to identify expected outcomes from a procedure and to develop a working environment where the staff works as a team, rather than as individuals, to reach that outcome.

- **Standard 4: Planning**   Plans for care delivery via the evaluation of systems. Promotes creativity and financial prudence in decision making. The Nurse Manager should develop plans for providing quality healthcare to patients, and also must consider the expense of healthcare.

- **Standard 5: Implementation**   Proposes the design for systems improvement to facilitate improved patient care. The plans for improving healthcare must be implemented effectively.

- **Standard 6: Evaluation**   Encourages the participation of others in the evaluation of planned change in patient care. Evaluates the implemented plan to determine what portions of the plan are effective, and then modify the ineffective portions of the plan.

And briefly, the nine Standards of Professional Performance are as follows:

- **Standard 7: Quality of Practice**
- **Standard 8: Professional Practice Evaluation**
- **Standard 9: Education**
- **Standard 10: Collegiality**
- **Standard 11: Collaboration**
- **Standard 12: Ethics**

- **Standard 13: Research**
- **Standard 14: Resource Utilization**
- **Standard 15: Leadership**

Emotional Intelligence

Emotional intelligence (EI) is a popular nurse management concept that focuses on managing emotions to effectively manage the department. The Hay Management Group, a leading management consultanting organization, defines EI as "the capacity for recognizing our own feelings and those of others, for motivating ourselves, and for managing emotions well in ourselves and others."

Nursing focuses on emotions: emotions of the nurse, the patient, and the patient's family and friends. The nurse must be compassionate and caring to everyone, regardless of the nurse's own emotions. Therefore, it is critical that the nurse be able to identify and manage his or her own emotions, which is the foundation of EI.

In addition to identifying and controlling emotion, the nurse must have empathy for others. That is, a nurse must be able to see a situation through the other person's eyes and emotions, and then modify his or her own emotions and management style to complement the other person's. Empathy is not sympathy. Empathy relies on a greater understanding of a person's emotional state. Sympathy is feeling sorry for a person without necessarily understanding the emotions behind their actions or behaviors.

As an example, suppose that a patient is overly demanding and antagonistic toward the staff. It is natural for the staff to react in kind and avoid the patient, resulting in the patient becoming increasingly frustrated. However, the situation can change dramatically if the Nurse Manager and the staff employ EI principles.

First, they must put themselves in the position of the patient and determine why the patient is overly demanding. One possible cause is that the patient is trying to control a situation that in reality they have little control over. Think for a moment. You're lying in bed practically naked sharing a room with a total stranger. You're sick and visitors for your roommate are coming in and out of the room throughout the day. You see your physician for a few minutes each day; other healthcare providers, some of whom you've never seen before, come in your room more often, poking and prodding you. You may not fully comprehend what is happening to you. In addition, you're concerned over your job and the cost of medical treatment.

Next, the Nurse Manager and staff must try to address the patient's underlying emotional concerns that are being expressed as being overly demanding and antagonistic. Instead of avoidance, they should become engaging and encourage the patient to verbalize his or her actual concerns so that the Nurse Manager and staff can rectify the situation.

Using EI helps the Nurse Manager do the following:

- Improve performance of the department by motivating the staff
- Increase the satisfaction of both the patient and the patient's family
- Improve relationships with other healthcare providers
- Develop a team that uses innovative approaches to problem solving

Summary

A Nurse Manager is given a goal and general guidelines by an upper-level manager, and then is expected to reach the goal by performing a number of tasks. The Nurse Manager must identify each task, determine how long each task takes to complete, determine the sequence in which tasks are performed, and determine resources (persons or things) needed to perform those tasks.

The Nurse Manager takes on several roles: planner, communicator, listener, decision maker, coordinator, delegator, supervisor, resources manager, staffing manager, clinician, evaluator, and leader.

A Nurse Manager is likely to adopt one of three management styles—authoritarian, democratic, or laissez-faire—or a combination thereof, depending on their experience, their level of confidence in managing others, and the capabilities and experience of the staff.

Leadership is a critical skill that a Nurse Manager must develop in order to influence others without necessarily having to rely on authority to direct their behavior. There are several leadership theories that a Nurse Manager can incorporate into their style of leadership: trait theory, behavioral theory, contingency theory, charismatic theory, transactional theory, transformational theory, and shared leadership theory.

The American Nurses Association established Standards of Practice and Standards of Professional Performance that guide the Nurse Manager through the process of managing a staff.

In the next chapter we'll focus on managing by taking a look at staffing and nursing care delivery models.

Quiz

1. "Establishes objectives and goals for each area and communicates them to the persons who are responsible for attaining them" is a management function described by:

(a) Peter Drucker

(b) Mills Training School

(c) Bellevue Hospital

(d) Florence Nightingale

2. What leadership style states that leaders are not born to lead but learn leadership behaviors?

(a) Behavioral theory

(b) Charismatic theory

(c) Transactional theory

(d) None of the above

3. Leadership is the ability to:

(a) Influence others through authority to direct their behavior

(b) Guide and influence others having the authority to direct their behavior

(c) Guide and influence others without necessarily having the authority to direct their behavior

(d) All of the above

4. A manager must identify tasks that are necessary to reach a goal.

(a) True

(b) False

5. The authoritarian style of management is one in which:

(a) The Nurse Manager provides little guidance to the staff.

(b) The staff participates in decision making.

(c) The Nurse Manager makes all the decisions.

(d) All of the above.

6. What leadership style states that the status quo changes when the staff adopts the leader's vision and beliefs for reaching a goal?

(a) Contingency theory

(b) Charismatic theory

(c) Transcultural theory

(d) Transactional theory

7. The Standards of Practice is a set of six standards developed by the American Nurses Association to help Nurse Managers and the nursing staff provide quality healthcare to patients.

   (a) True

   (b) False

8. What leadership style states that leadership behaviors are dictated by a situation?

   (a) Charismatic theory

   (b) Contingency theory

   (c) Transactional theory

   (d) None of the above

9. Emotional intelligence (EI) is a popular nurse management concept that focuses on managing emotions.

   (a) True

   (b) False

10. Listening provides the opportunity to receive valuable feedback that is used to avert some problems and resolve others.

   (a) True

   (b) False

CHAPTER 3

Nursing Care Delivery Models and Staffing

Imagine for a moment that you are a Nurse Manager who is assigned a staff of 20 nurses, 8 nursing assistants, and 4 unit clerks to help care for 35 medical-surgical patients 24 hours a day, 7 days a week. How will you plan to care for these patients?

Your challenge is to nurse patients back to health without directly providing the care yourself. You do this by using techniques—and a few tricks—that Nurse Managers use every day to successfully provide care for their patients.

A Nurse Manager devises a game plan to care for patients and then coaches the staff to carry out the plan. The Nurse Manager first sets a goal, and then establishes guidelines for reaching the goal, after which focus turns to managing the staff as they work toward the goal.

A key element of every game plan is a *nursing care delivery model*, which defines a framework within which to care for patients. In this chapter, you'll learn how to develop your own game plan around a nursing care model and how to use proven professional techniques to manage your staff members as they care for your patients.

What Is a Nursing Care Delivery Model?

The question remains, how will you plan to care for these patients? Rather than creating a plan from scratch, a Nurse Manager adopts a nursing care delivery model. A model describes a way of doing something. In this case, the model describes how to deliver nursing care to patients.

Unlike a script that specifies the details of what needs to be done, a nursing care delivery model provides a structure within which the Nurse Manager and the staff are free to develop details for caring for each patient. This structure is referred to as a *framework*.

For example, a nursing care delivery model might specify that one member of the staff will set up an I.V. for every patient who requires one, but does not say how and when to set up an I.V. Those details are left to the healthcare facility or the Nurse Manager.

A different nursing care delivery model might require that each nurse be assigned to care for seven patients. The nurse provides all the care for each patient using procedures specified by the healthcare facility or the Nurse Manager.

Nursing care delivery models define a framework for providing a cost-effective, practical way to care for patients using limited resources. It strikes a balance between the cost of care and quality outcomes. *Cost* refers to the expense of delivering patient care based on the definition of the model. *Quality outcomes* refers to the quality of care that patients receive from the staff.

For example, a patient receives the best possible care if a team of healthcare professionals devotes its time to caring only for that patient. Very few patients receive such care because the cost is prohibitive. In comparison, a nurse might be assigned to care for 35 patients in a medical-surgical unit. The cost for this care may be very reasonable, but the quality of care is substandard. It is impossible for one nurse to deliver quality care to 35 patients.

Nursing care delivery models are evidence-based and consider the changing demands of patient care, advancements in technology, healthcare reimbursement practices, available staffing, and patients' expectations. Researchers set out to develop a nursing care delivery model for typical situations that a healthcare provider might encounter. However, in practice, Nurse Managers blend together different nursing care delivery models to create a model that works for a particular healthcare facility.

Let's take a look at popular nursing care delivery models.

CASE MODEL

The case model is one of the first nursing care delivery models and was introduced in the 1800s to provide a framework for caring for sick family members of working families. During this period, it was customary for the sick to be cared for at home by family members rather than in a hospital.

Home care became a burden for many families whose members were unable to give care because they worked outside the home. The case model addressed this situation by specifying that a nurse care for patients in their homes. Each patient is considered a *case* and all the cases assigned to a nurse collectively are called a *caseload*.

The case model is similar to today's private-duty nursing, where a nurse cares for one patient. Nurses in the 1800s who used the case model became live-in caregivers. The case model faded from popularity as live-in nurses became hard to find and more sickly patients were cared for in a hospital setting rather than at home.

FUNCTIONAL NURSING MODEL

A nursing shortage developed as the world went to war in the 1940s. It was during this time that the functional nursing delivery model was introduced. The functional nursing model specifies that one caregiver performs the same procedure on every patient who requires the procedure. For example, one caregiver will start an I.V. for every patient who is scheduled to receive one.

It was during this period that state governments began to authorize licensed practical nurses and nursing assistants to compensate for the shortage of registered nurses. The functional nursing model worked well because there were groups of healthcare providers with different skill levels. For example, a nursing assistant would provide morning and night care for a group of patients, freeing a registered nurse or licensed practical nurse to administer medication to these patients.

The advantage of the functional nursing model is that each healthcare provider becomes proficient in the procedures he or she performs. The drawback is that no one views the whole patient. Each healthcare provider views the patient only in relation to the procedure that is being performed.

Furthermore, the functional nursing model tends to confuse the patient. The patient sees a number of caregivers each shift and feels a lack of continuity in the care. There isn't one caregiver who seems responsible for the entire care of the patient. And to add to the confusion, the patient isn't sure who to speak to about his or her care.

The functional nursing model works well in situations where the Nurse Manager personally directs relatively inexperienced or per diem healthcare providers to care for a group of patients. Each healthcare provider knows how to perform one or more procedures, but isn't responsible for the patients' overall care plan. Few healthcare facilities adopt this model, although ambulatory care facilities that specialize in day surgery use a modified version of this. In these facilities, one nurse pre-ops the patient, one nurse takes the patient to surgery, and one nurse recovers the patient.

TEAM NURSING MODEL

Many weaknesses in the functional nursing model were addressed with the introduction of the team nursing model in the 1950s. The team nursing model uses a team of healthcare providers to care for a patient; the team is a discrete unit led by the Nurse Manager. Healthcare providers are permanent members of the team that work together to give the patient total care.

There are three advantages to using the team nursing model:

- Each team member feels responsible for total patient care. Thus, there is an open and continuous line of communication among team members.
- The patient becomes familiar with the team, because the patient interacts with the same group of healthcare professionals every day.
- Assignments are based on each team member's level of education. For example, a nursing assistant takes care of morning and night care while the registered nurse assesses the patient's condition.

There are also disadvantages to this model:

- The team spends time discussing the patient's progress. These meetings often take longer than is necessary.
- Schedule changes among the team compromise the continuity of care. A team member who is away for a day must be provided an update on the patient's condition.
- Resentment can grow among team members if one or more team member perceives that they are always assigned to unpleasant tasks because of education or license, or because the Nurse Manager favors one member over another.
- The Nurse Manager must have excellent communication skills and a knack for resolving conflicts among team members.

The team nursing model requires that the Nurse Manager become an active team leader and use interpersonal skills to keep up the team spirit during challenging times. The Nurse Manager who adopts a transformational leadership style (see Chapter 2) will likely find success using the team nursing model, because that style helps the Nurse Manager to demonstrate integrity and to earn the team's trust. The Nurse Manager conveys a vision and beliefs for reaching a goal, and then empowers the staff to work toward that goal.

Some version of the team nursing model is commonly used in hospitals and long-term care facilities.

PRIMARY NURSING MODEL

The primary nursing model, introduced in the 1960s, states that one nurse is responsible for the total care of a group of patients. The major objective of this model is to place the registered nurse back at the patient's bedside; whereas the team nursing model has the registered nurse share patient care with team members, relieving the registered nurse from many bedside tasks.

Under the primary nursing model, the primary nurse is responsible for all aspects of patient care while the patient remains on the unit. This supports the philosophy of nursing that states nursing is a knowledge-based profession and not a task-oriented skill. That is, the nurse must assess, analyze, plan, implement, and evaluate when caring for a patient, rather than simply performing procedures ordered by a physician.

The following are the advantages of the primary nursing model:

- Responsibility for patient care is decentralized from the Nurse Manager to the primary nurse.

- There is one nurse who is responsible for care of the patient 24 hours a day while the patient is in the unit. Associate nurses assume primary care when the primary nurse is unavailable (days off, for example).

- It is clear to everyone who is in charge of patient care, which tends to increase satisfaction of the patient, the patient's family, the patient's physician, and the primary nurse.

- The primary nurse is the coordinator of the patient's care among other units in the hospital. The primary nurse identifies and resolves conflicts regarding tests, medication, and procedures that the patient receives.

- The primary nurse is recognized as a professional.

There are disadvantages to the primary nursing model, too:

- In reality, one nurse cannot be responsible for the patient's care 24 hours a day, 7 days a week.

- The model is costly, because healthcare facilities perceive that a registered nurse must provide all patient care.

- The primary nurse cannot be responsible for care given outside the unit, such as when the patient is transferred for tests.

- The primary nurse and associate nurses may not fully communicate with each other when they hand off the patient.

The primary nursing model works well for Nurse Managers who adopt a democratic and transactional style of management, because primary nurses work autonomously.

However, this model fell out of favor because of the nursing shortage in the 1970s and 1980s, and is in limited use today for the same reason.

THE CASE MANAGEMENT NURSING MODEL

The case management nursing model is designed to move the patient through the healthcare system in the most cost-efficient manner without sacrificing the quality of patient care. The case management model is based on a nursing care plan (sometimes referred to as a multidisciplinary action plan, or MAP) that specifies what is to be accomplished each day the patient is in the hospital according to the patient's diagnosis.

A nurse case manager (CM) manages the patient through the treatment process described in the case management model. The treatment process includes in-hospital treatment, home treatment, and treatment in other facilities (such as rehabilitation and long-term care facilities). The Nurse Manager also coordinates treatment with insurers to ensure that reimbursements are forthcoming. Any deviation from the plan is documented, and efforts are made to get the patient back on the plan. Variances are then addressed when the plan is revised.

A case manager doesn't provide patient care, but instead coordinates other healthcare providers in caring for the patient. Some healthcare facilities have unit-based case managers or disease management case managers. A unit-based case manager manages the patient while on the unit, whereas a disease management case manager manages the patient with a particular disease, regardless of the patient's location within the hospital.

Numerous hospitals are using the case management model of care, and the number is growing. The greatest advantage to the hospital is that the close attention paid to a patient by one CM, who closely follows the patient's progress through MAPs, reduces the patient's length of stay. Hospital reimbursement is a fixed payment, based on Diagnosis-Related Groups (DRGs) and a standard length of stay. The sooner the patient leaves the hospital, the fewer resources the patient will consume, and the more likely the hospital will make money. By following a precise plan, patients also receive quality care.

Patients and families appreciate having one person to talk with who knows all aspects of the patient's hospitalization, from the physician orders, to scheduled tests, to discharge date, to the arrangement of any home care needs.

The only disadvantage arises when the model is poorly implemented. If the hospital adds too many new CM positions, then the model is costly and the hospital will not garner any savings. If the departments of Utilization Management, Discharge Planning, and Social Services do not participate, many tasks will overlap and even be redundant. Successful hospitals create a Department of Case Management, and reduce the number of staff by cross-training RNs from the above-mentioned departments to be case managers. Social workers continue to utilize their special knowledge, but fewer social workers are needed. This is not an easily implemented change, but it is a necessary one in order to reap all the benefits of the Case Management nursing model.

The disease management model is more frequently seen in community health facilities, where disease management case managers work to keep patients out of the hospital. Disease management focuses on one disease entity, perhaps diabetes or congestive heart failure, and uses evidence-based clinical guidelines to improve patient outcomes. For nurses to be successful in managing specific diseases, they must have a thorough understanding of the course of the disease. They must be able to locate the patients in the hospital who will benefit from education regarding the disease, and who will likely be compliant with guidelines to control the disease. Disease management is advantageous for the hospital because it reduces a patient's length of stay, and it is also advantageous for the patient because it helps them to get home faster, and may prevent worsening of the disease and future hospitalizations.

TOTAL NURSING CARE MODEL

Today, this model is most frequently used in critical care units, such as Intensive Care and Labor & Delivery. It is similar to the primary care model, in that the nurse assumes total care for the patient (often just one or two patients during a shift). The nurse is not responsible after his or her shift ends, there are no "associate" nurses, and the patient is not followed to another unit. With one nurse being totally accountable for care, there is better communication with colleagues from other disciplines/departments and with family members. A disadvantage of this model is that the professional nurse is doing certain tasks that could be delegated to an ancillary staff person.

Staffing

Staffing is the process of having the appropriate number and mix of healthcare providers available to care for the actual or projected number of patients to achieve cost-effective, quality patient care.

A nursing care delivery model doesn't tell a Nurse Manager the number of nurses needed to cover their unit. The model simply gives them a framework within which to analyze their needs in order to set their staffing budget. A staffing budget for a unit is defined in full-time equivalents. A *full-time equivalent (FTE)* is the total number of full-time hours of workers. A full-time employee works eight hours per shift.

Two part-time employees each work four hours per shift, which is equivalent to a full-time employee working one shift. On hospital personnel reports, the number 1 is an FTE and the fraction 0.5 is one-half an FTE.

Let's say that the unit has 7 FTEs working the day shift. The Nurse Manager has created the schedule as 3.5 RNs, 1.5 LPNs, and 2 Unlicensed Nursing Personnel. This equates to three RNs working eight hours each, one RN working four hours (this is a surgical unit, so the four hours are 7 A.M. to 11 A.M., a busy time getting patients ready for the operating room), one LPN working eight hours, and one LPN working four hours, maybe from 11 A.M. to 3 P.M. to cover other staff members for lunch break time. This example highlights the flexibility a Nurse Manager has in allocating staff. It is often more useful to have two part-time employees than to have one person filling the full eight-hour shift.

Budgeted positions are defined as FTEs, rather than a specific number of employees. This gives the Nurse Manager flexibility when creating the master staffing plan. Two part-time employees (noted as .5 FTE twice on the budget) may be hired, rather than one full-time FTE. The part-time employees may then be assigned to shifts that require some supplemental help, but not to the extent of requiring a full-time person. The Nurse Manager is expected to stay within the staffing budget for the unit on a monthly or quarterly basis, depending on the policy of the healthcare facility. This means that the unit can exceed a daily budget if patient activity is high, as long as this increase is offset by a decrease in hours at some point during the month or quarter. If the number of patients or the patient *acuity* (level of care) increases, the Nurse Manager will have a valid justification for being over budget for the time period in question.

Staffing requirements are determined by applying a formula that considers the patient's acuity and the length of time that the care is to be given. At the beginning of a shift or every four hours, depending on hospital policy, a nurse assesses the level of care for each patient and enters that information into the patient classification system.

Patients are assessed according to tasks that are associated with their care requirements. The patient classification system determines the number of hours required to perform each task based on the complexity of the task. For example, a burn patient may require 24 hours of care each day if the burns are severe, while a total hip replacement patient may require only 3 hours of care per day. The system then forecasts the number of FTEs required to care for the patient, recorded as *hours per patient day (HPPD)*. This forecast reflects the patient's current level of care, which can differ from one shift to another as the patient's condition changes. In order to arrive at the staffing budget, the Nurse Manager tallies the number of hours, averaged over a period of time, for each patient to arrive at the staffing budget using the average daily census.

Some healthcare facilities have a central staffing office that collects data from patient classification system forms from all the units, and attempts to allocate staff in the most appropriate manner. Personnel in this office know the census and the acuity of patients on each unit, along with having the list of who is working on each unit for the coming shift. Nurses may be reassigned based on this knowledge. Other facilities leave the allocation of day-shift staff to a group of Nurse Managers who cover similar units (medical, surgical, psychiatric, etc.), and these Nurse Managers reallocate staff based on the same data and a more personalized look at the units. The divisional supervisors are responsible for the allocation on the evening and night shifts.

How does a Nurse Manager decide how many employees are required for a particular patient care unit? Once each year the Nurse Manager must propose a personnel budget for the coming year, determined in the following manner: The Nurse Manager receives a report from the data management or finance department. This report shows, for example, that on the 25-bed unit the average daily census for the year was 19 patients. The patient classification system reports that the typical patient required five hours of care each day. So, 19 patients times 5 hours of care results in the unit requiring 95 care hours each day.

If the shift is eight hours long, the Nurse Manager divides 95 by 8 to determine that the unit requires 11.7 FTEs (rounded up to 12) to staff the unit for 24 hours each day. An FTE is one person working 40 hours a week or two people each working 20 hours. Since each person is off two days a week, working 40 hours and off 16 hours, the number of FTEs is multiplied by 1.4 ($16 \div 40 = 0.40$), so the unit needs 16.8 FTEs, rounded up to 17, as a basic staffing number. Vacation, holiday, personal, and sick days off may be covered by per diem or float pool personnel. If the unit is to cover these days off, an additional percent of staff must be added. The actual percent will depend on how many benefit days (days one is paid for but does not work) the hospital policy or union contract allows.

Now the Nurse Manager must decide how to categorize the 17 FTEs—that is, how many RNs, LPNs, and nonlicensed nursing assistants need to be present. The Nurse Manager and unit secretaries do not count in this equation, as they are not direct care givers. A prototypical mix, rounding the numbers up or down, is 60 percent RN (17 FTEs $\times$ .60 = 10.2) that will be 10 RNs, 25 percent LPN ($17 \times .2 = 4.2$) that will be 4 LPNs, and 15 percent nursing assistants ($17 \times .15 = 2.5$) that will be 3 nursing assistants. How many of each category should work days, evenings, and nights? A hospital will decide this based on how busy the unit is on each shift. In the example shown in Table 3-1, the hospital wants the unit FTE allocation to be approximately 50 percent of staff on days, 35 percent on evenings, and 15 percent on nights.

Table 3-1   A Work Schedule

Category	Day	Evening	Night	Total
RN	5.0	3.0	2.0	10.0
LPN	2.0	2.0	0.0	4.0
Aide	1.0	1.0	1.0	3.0
Total	8.0	6.0	3.0	17.0

There are many factors that influence staffing besides those mentioned in this section. These include union contracts, configuration of the unit (long hallway vs. pods), timing of admissions, transfers, and discharges, and expectations of management, physicians, and the community.

STAFFING CHALLENGES

In an ideal world, everyone shows up on time for his or her shift. In the real world, this doesn't happen because people get sick or are otherwise unable to report to the unit. When this occurs, the Nurse Manager must find someone to take over the shift. Here are a few ways to handle this problem:

- Request someone from the per diem pool.
- Ask for a volunteer from the current shift to work the next shift.
- Ask for a volunteer from the current shift to work part of the next shift and ask a member of the third shift to report to the unit early.
- Ask the nursing director to float someone from a different unit for all or part of the shift.
- Ask a person who has a day off to come to work to cover the shift.
- Call in an agency nurse, if that is an option in the organization.

If the problem persists due to excessive vacancies, the organization may employ traveling nurses for assignments as long as 6 months. In extreme situations, the Nurse Manager may have to work to cover the shortfall.

RECRUITMENT AND RETENTION

The quality of patient care is dependent on the skills of the staff and the support the staff receives from management and other areas of the healthcare facility. The nursing department of the healthcare facility is responsible for recruiting candidates, who then undergo an interview process to determine if they should be invited to join the staff.

Depending on the healthcare facility, the Nurse Manager may play a key role in the hiring process. In some healthcare facilities, the Nurse Manager is responsible for selecting staff for the unit from a pool of candidates sent by the Human Resources department. These candidates might be employees looking to transfer to a different unit. Other candidates might not be employed by the healthcare facility, but apply for a position in response to advertising.

In either case, the Nurse Manager must evaluate each candidate in the same way during the interview process. The Nurse Manager should begin the interview with the candidate by stating that the objective of the interview is for both the candidate and the manager to decide if the position is right for the candidate. Notice we didn't say "is the candidate right for the position." This subtle difference signals to the candidate that the Nurse Manager is supportive and has the candidate's best interests at heart. If the candidate is not offered the position, the candidate should understand that his or her current skills do not fulfill the job requirements, rather than feel that they have failed personally.

The Interview

The Nurse Manager's interview with the candidate focuses on whether or not the candidate will be a good team member, rather than on the candidate's caregiving skills. In many healthcare facilities, the nursing department determines through interviews and testing whether a candidate has the proper caregiving skills. Therefore, the Nurse Manager should concentrate on whether or not the candidate will work well with the current staff.

The Nurse Manager may request several staff members to interview the candidate in an informal manner while showing the person around the unit. Several hypothetical questions may be posed to the candidate to determine the fit with the current team. As an example, the Nurse Manager may inquire how the candidate handled, or would handle, an interpersonal conflict with a colleague. Will the candidate immediately go to the Nurse Manager for a resolution, try to work out the problem one-to-one with the colleague, or simply ignore the situation and avoid the colleague in the future? Using her or his own leadership style, the Nurse Manager then can determine if the answer is compatible with the Nurse Manager's expectations and the unit's culture and philosophy. The term "fit" simply means that the candidate's personality and work ethic blend with those of the current staff.

Let's say that you are a Nurse Manager and your current staff is quite competent and willing to help each other. Everyone treats each other respectfully. A person with the same temperament will fit well with the group. However, a candidate who has a "superstar" attitude and considers those with less skill as inferior is bound to clash with the staff and would not be a good fit. Ask the candidate why they are leaving their current position. If the candidate says the new position offers a better opportunity, then follow up and ask how they arrived at that conclusion. If the candidate alludes to a poor working environment in their current position, then ask the candidate to describe those working conditions.

Your objective is to identify problems the candidate may have with his or her current employment, and then decide if the same problems might arise in your unit. If so, then the position is not right for the candidate.

Be aware that certain questions cannot be posed to a prospective employee. Such questions include age, sexual preferences, and religious beliefs. Human Resources (HR) personnel can provide guidance on this aspect of interviewing.

By the end of the interview, the Nurse Manager should have gathered sufficient information to decide whether or not the candidate is a good fit with the staff. The Nurse Manager should conclude the interview in a positive but neutral manner.

Hiring

In some healthcare facilities, the Nurse Manager, in conjunction with the HR department, decides which candidate is offered the position. The HR department, after checking the candidate's license, work history, education, and references, makes the actual offer. The offer is made in writing according to the policies of the healthcare facility. Technically, the offer is not a contract because the offer doesn't specify duration of employment, which is a requirement of a contact. However, the offer binds the healthcare facility to certain obligations that are set forth in state and federal law. These obligations are beyond the scope of this book, and typically are not the concern of the Nurse Manager.

A new employee goes through a probation period during which time the employee demonstrates his or her competence to perform the requirements of the position. The probation period begins with an orientation that introduces the new employee to the policies and procedures of the healthcare facility and to the unit.

Depending on the healthcare facility, the nursing office or HR may direct the orientation process without involving the Nurse Manager. The new employee reports to the unit only after the general orientation is completed. Then a unit-based orientation begins, to acquaint the person with specific procedures related to the specialty of the unit.

The Nurse Manager must adopt the trust-but-verify approach to working with a new employee. That is, trust that the employee is a competent healthcare provider, but verify this. They should assign the new employee to work with another staff member, and ask that staff member to oversee the new employee's work—that is, verify that the work is performed properly.

Termination

It is customary that an employee on probation can be terminated for any reason and has no recourse to appeal the dismissal. At the conclusion of the probationary period, an employee can be terminated only after the Nurse Manager follows the policies of the healthcare facility.

An employee cannot be terminated at will. Employees are protected by the National Labor Relations Act, the Civil Rights Act of 1964, and by policies within the healthcare facility. An employee can be terminated for cause. Therefore, the Nurse Manager must establish cause before terminating an employee. This typically means that the employee violated a policy of the healthcare facility or violated the law. The Nurse Manager should consult with the facility's HR department to determine the healthcare facility's definition of cause.

In many healthcare facilities, the Nurse Manager has authority to temporarily reassign an employee or even send the employee home, with pay, if the employee makes a grievous error that jeopardizes the well-being of a patient or another employee. The Nurse Manager then has time to consult the nursing office or HR on how to proceed.

All efforts should be taken to avoid terminating an employee. If the employee is engaging in unsuitable behavior, then the Nurse Manager should discuss the situation with the employee, give advice on how to correct the behavior, and give them the opportunity to do so. Most employees try to do a good job and appreciate the opportunity to correct missteps. Before making formal written reprimands that are placed in the employee's personnel file, the Nurse Manager should issue a verbal warning and provide guidance or education to correct the unacceptable behavior.

The Nurse Manager should set expectations, with input from the employee, and then compare the employee's performance against the expectations. If the employee fails to meet these expectations, the Nurse Manager should explore reasons for this failure with the employee, and together come up with a plan for addressing the situation. If expectations still are not being met, the Nurse Manager should ask the employee if they feel the position is right for them. The Nurse Manger may acknowledge concerns that the employee might be feeling, even if those concerns are not voiced, and help the employee think through their situation by exploring viable options.

The Nurse Manager may suggest helping the employee explore other positions within the healthcare facility. This implies that the employee's difficulties meeting expectations of the current position won't prevent the employee from finding a more suitable position within the organization. This, of course, applies if the employee is unable to meet the requirements of the specialty area. It does not apply if the employee is constantly late, abuses sick time, or receives quality of care complaints from patients.

The Nurse Manager may also suggest that the facility will work with the employee to find a position in a different healthcare facility by postponing the start of the formal termination process for three months. This gives the employee the opportunity to find a more suitable position and then resign. It is always easier to find new employment when you are currently employed.

The Nurse Manager should keep the nursing and HR departments informed of his or her concerns and approach to resolving the issues, without official termination. The Nurse Manager should proceed only if the nursing and HR departments concur with their approach to handling the situation, because it may be that a less formal approach might expose the Nurse Manager and the healthcare facility to legal action.

Discipline

Probably the most disliked responsibility of a Nurse Manager is to discipline an employee (no longer on probation) for inappropriate behavior, because in many situations the Nurse Manager has developed a working and sometimes personal relationship with employees. Disciplining an employee jeopardizes this relationship and, in rare situations, exposes the Nurse Manager to retaliation and physical abuse. Furthermore, no one wants to be responsible for placing a colleague's financial well-being at risk. However, the Nurse Manager's primary job is to provide safe, quality healthcare to patients, and they must place patient care above maintaining a friendly relationship with employees. At times the Nurse Manager will have to correct an employee's behavior to assure quality patient care.

Disciplining an employee involves risk, because the Nurse Manager must be able to prove that the employee's conduct is inappropriate. To do so, the Nurse Manager must thoroughly investigate the situation and develop and document supportive evidence that would lead a reasonable person to believe that the employee displayed the conduct; and that the conduct is counter to the healthcare facility's policy, acceptable nursing practice, or the law.

Before trying to correct the behavior, the Nurse Manager should do the following:

- Have a good understanding of the healthcare facility's rules, regulations, and policies on discipline. Consult with the HR department before taking any action against the employee, except in extreme circumstances.

- Anticipate that their claims will be challenged by the employee and investigated by third parties, including legal counsel for the healthcare facility, the employee's attorney, and possibly state and federal government regulators.

- Document events that led up to the disciplinary action and the steps that the Nurse Manager and others in the healthcare facility took to correct the problem. The documentation must be in writing, and clearly state the date, time, nature of the event, personal observations, and observations made by other staff members. Documentation provided during orientation, and proof that the employee received information concerning how to act as a responsible employee should be included.

- Contact the HR department before confronting the employee. This ensures that the Nurse Manager follows all the disciplinary rules.

- Meet with the employee in private. The Nurse Manager should explain their concerns and listen to the employee's response before taking any disciplinary action. They may discover that the employee took proper action given the circumstance.

- Apply the healthcare facility's appropriate discipline policies. Typically, the employee is given a verbal warning, followed by a written warning, suspension, and then termination if the behavior isn't corrected. The penalty for more severe infractions, such as falsifying medical records or theft of property, usually is immediate termination.

It is very important that the Nurse Manager be consistent when taking disciplinary action. Any inconsistency will cause ill feelings among other employees and can be grounds for an employee to bring legal action against the Nurse Manager and the healthcare facility for discrimination.

Retention

Maintaining a comfortable work environment goes a long way toward retaining a quality staff, which should be one of the Nurse Manager's most important goals, second only to providing quality healthcare for patients. It is the Nurse Manager's responsibility to create a comfortable and safe work environment by ensuring that the staff is treated with respect and dignity, and given the support needed to properly care for their patients.

It is important that the Nurse Manager and staff realize that nursing is a task-driven, autonomy-restricted work environment. Policies and procedures that are created by a committee consisting of all levels of professional nurses in the organization define nursing practice in your facility. Nurses follow orders written by physicians, and seldom work autonomously. Staff nurses are hourly employees, and are entitled to overtime pay. Nurses in management positions are salaried and do not receive overtime pay.

Creating a comfortable work environment is challenging under these conditions:

- Nurses feel their pay is low relative to the life-and-death responsibilities that they accept daily.
- There is a perception of poor staffing, with nurses caring for too many patients.
- Nurses are frequently required to float to units where they lack orientation and training, exposing patients to less than quality care and nurses to making errors.
- Frequent and sometimes mandatory requests are made to work overtime and to rotate shifts.
- Nurses are asked to perform nonnursing duties.
- Support from other departments is lacking.
- Nurse Managers and administrators adopt an autocratic management style that frowns upon creative solutions to healthcare problems.

These challenges are inherent in the nursing profession, and most nurses and staff realize this. As a Nurse Manager, you won't be able to eliminate these problems, but you can take steps to minimize their effects on your staff. Here's what a Nurse Manager can do to create a comfortable work environment:

- Establish a shared governance model of management on the unit. This may include having staff members prepare the timesheet (work schedule), having staff participate in decisions affecting unit work routines, requesting that staff form committees to revise and rewrite policies, and being a Nurse Manager who listens to staff before making unit changes.
- Work to create a team that includes everyone on the unit, not just the nurses.
- Take a leadership role in establishing good working relationships with other nursing units and other hospital departments.
- Be flexible. Interpret a policy to permit more staff to be off on a holiday than previously allowed. Be flexible with vacation and time off when possible. If 7 A.M. to 3 P.M. is difficult for one person, consider allowing that person to work 8 A.M. to 4 P.M. Consider the concept of job sharing, if that will retain one or two excellent employees.
- Celebrate and praise. Everyone responds to praise for a job well done. Praise twice as frequently as you admonish. Celebrate higher than average patient satisfaction scores, staff members getting an advanced degree, and

anything else that warrants it. Depending on the personality and leadership style of the Nurse Manager, this may be easy or more difficult.

- Be fair to all.
- Be honest. Don't try to "sugar coat" bad news, and don't over inflate good news.
- Respect confidentiality in all things.
- Have an "open door" policy so staff members may discuss any issue with you privately.
- Use "we" more often than "I."
- Act as a role model for everyone.

Based on your own nursing experience, you can surely add to this list.

Union and Union Contracts

Staff members in some healthcare institutions have formed a bargaining unit commonly referred to as a *union*. The union negotiates terms of employment with the healthcare facility for members of the union. These terms are contained in a *collective bargaining agreement*, which is a contract between the union and the healthcare facility.

The collective bargaining agreement specifies conditions under which the staff provides healthcare to patients. Both the staff and the Nurse Manager must adhere to these conditions; otherwise, they are in violation of the agreement. In essence, the bargaining agreement spells out what management and union employees can and cannot do. For example, the bargaining agreement may say that nurses cannot be required to work mandatory overtime, or that they be given two weeks' notice before rotating to a different shift.

Neither the Nurse Manager nor a staff member has the right to alter the terms of the agreement. The healthcare facility and representatives of the union must formally agree to any changes to the agreement. They may, however, have a clause stating that under certain conditions, the company and union can have a non-precedent-setting agreement in which the terms can be altered for a specific individual or incident. For example, a hospital may currently offer nurses an option to work 24 hours every weekend of the year (two 12-hour shifts) and receive pay for 40 hours worked each week. If the Nurse Manager notes that this option causes the unit to be over budget, the Nurse Manager may want to discontinue the practice. The Nurse Manager cannot do this. The hospital must open formal contract talks with the union to seek such a change.

MANAGING UNION PERSONNEL

Managing a unionized staff can be challenging for a Nurse Manager, because both sides must follow rules defined in the bargaining agreement. The Nurse Manager cannot simply use good management judgment when making staff decisions. Instead, the Nurse Manager must consult the bargaining agreement to determine if the decision conforms to the agreement.

A Nurse Manager can still have a cordial working relationship with unionized staff, as long as they complies with the terms of the agreement. The HR department will provide the Nurse Manager with a copy of the agreement that governs all agreed-upon aspects of managing the nursing department. The agreement will contain policies and procedures for the more common interactions that the Nurse Manager has with the staff.

The Nurse Manager must also consult with HR before disciplining unionized staff, to be sure that the agreement permits such action. Some agreements have a three-step positive, progressive disciplinary process. Depending on the alleged infraction, the employee may be placed on any one of the three steps, with step 1 being the least severe, and step 3 being the final step before termination.

A bargaining agreement specifies how alleged violations of the agreement are handled. Typically, this involves a formal hearing where a union representative, possibly an attorney, represents the staff member. The decision of the HR department's hearing officer can be appealed to an independent third party, called an *arbitrator*, in a process called *binding arbitration*, which means both the healthcare facility and the union agree to abide by the arbitrator's decision.

The bargaining agreement dictates how union members are managed. However, this agreement doesn't cover every situation, simply because negotiators cannot think of every situation that could arise. This means the Nurse Manager will likely encounter situations with the unionized staff that aren't covered by the bargaining agreement. In these cases, a Nurse Manager cannot simply make up policy, as they might with nonunion employees. Instead, the Nurse Manager and the employee may find a solution together, or the Nurse Manager will need to raise the issue—and suggest a solution—with the nursing office or HR, and ask that the healthcare facility representative and the union representative resolve the issue.

There are two ways in which issues not covered by the bargaining agreement can be resolved. Both sides can agree to an addendum to the agreement. An *addendum* is an add-on to the original agreement that occurs during the contract period. The other approach is to hold the issue over to when the agreement is renegotiated. The healthcare facility representative and the union representative collectively decide which of these approaches is taken to resolve the issue. Typically, minor and urgent issues are resolved through an addendum, whereas major issues usually are held over until renegotiation of the agreement. The healthcare facility representative must weigh the importance of resolving an issue against the risk of opening negotiations with the union, which allows other issues to be placed on the table and may delay a resolution.

The Nurse Manager is not involved at the table in negotiations. The Nurse Manager should focus on providing quality healthcare to patients and creating the best possible work environment to accomplish this. Before the healthcare organization and union go into renegotiation of a contract, each side (privately, of course) makes a list of what it hopes to achieve. The union will concentrate on higher pay or better working conditions, such as increasing the number of staff nurses in the hospital in order to provide better care or more time off. The healthcare facility will review the current contract and seek input from all management nurses, including Nurse Managers, relating to articles in the agreement that have proven problematic to running the department. This might be, for example, an article that states that any nurse with 20 or more years of experience at the hospital must be given every weekend off. Units with several nurses in this category may have trouble staffing on the weekends. At the table, the hospital negotiators will try to get that amended to 25 years, or have the article be dismissed altogether. The hospital may also want to do away with a 12-hour-shift option, if that has caused financial difficulties or resulted in more medication errors committed during the last 2 hours of that long shift.

A bargaining agreement is just that—bargaining. "If you provide this, I will provide that." Everyone works toward a win-win situation, and since so many contracts are ratified each year, it is safe to assume that the same situation is achieved in hospital-union negotiations.

GRIEVANCE PROCEDURE

A union member has the right to challenge behavior by management that the union member considers a violation of the bargaining agreement. The challenge is called a *grievance* and the process of resolving the challenge is called the *grievance procedure*. Think of a grievance as a complaint, and the grievance procedure as the way the union member gets his or her complaint heard by an authority that can resolve the issue. A grievance may have merit or be without merit; regardless, the grievance must be heard and resolved according to the grievance procedure.

The grievance procedure is specified in the bargaining agreement and typically involves several steps:

1. The union member brings the grievance to the attention of the Nurse Manager by presenting facts of the situation and the specific clause in the bargaining agreement that is being violated. The Nurse Manager reviews the facts and the bargaining agreement and then determines if a violation has occurred. If so, then the Nurse Manager corrects the situation. If not,

then the Nurse Manager notifies the union member, usually in writing, that the grievance has been denied.

2. The union member brings the grievance to the union representative, who then brings the grievance, accompanied by the staff nurse, to the nursing supervisor. The union representative presents the grievance and the nursing supervisor reaches a decision—the situation is corrected or the grievance is denied.

3. The union representative brings the grievance to the chief nursing officer of the healthcare facility, who then weighs the facts and offers suggestions as to how to correct the situation or denies the grievance.

4. The union representative brings the grievance to the HR director, who convenes a hearing panel that hears the grievance and makes a ruling.

5. If the hearing panel denies the grievance, the union may have the right to appeal the grievance to binding arbitration or mediation, depending on the terms of the bargaining agreement. *Mediation* is a process where an independent third party, called a *mediator*, investigates the grievance, assembles what the mediator believes to be the facts, and tries to bring about an agreement between both parties.

The decisions of the Nurse Manager can be challenged (grieved) by a member of the unionized staff who feels that the decision violates the bargaining agreement. Therefore, the Nurse Manager must be prepared to defend the decision as abiding by the bargaining agreement.

PAST PRACTICE

Work rules are specified in the bargaining agreement. Rules not included in the bargaining agreement are unenforceable with one exception, called *past practice*. A past practice is something that management did in the past, but that is not included in the bargaining agreement. Past practices are long-standing practices that have been accepted by both parties during that period. They are not written, but they are binding.

For example, suppose the organization has always provided employees a paycheck every two weeks. The organization now wants to go to a once-a-month paycheck plan. The union can declare this paycheck schedule a past practice and require management to continue the practice, even if it is not mentioned specifically in the agreement. Past practices can only be stopped if both parties agree during a collective bargaining session.

It is important to understand that a Nurse Manager can inadvertently create a past practice by giving union members privileges that are not covered in the bargaining agreement. Any disagreement about a past practice is resolved through the grievance process.

Summary

A nursing care delivery model provides a framework for delivering nursing care to patients. There are several popular nursing care delivery models.

The case model was one of the first nursing care delivery models and provided a framework for a nurse to provide home nursing care to one or more patients. The functional nursing model specifies that one caregiver performs the same procedure on every patient who requires the procedure. The team nursing model uses a team of healthcare providers to care for a patient under the direction of a Nurse Manager. The primary nursing model states that one nurse is responsible for the total care of a group of patients. The case management nursing model is designed to have a case manager move the patient through the healthcare system in the most cost-efficient manner without sacrificing the quality of patient care.

A nursing care delivery model doesn't identify the number of nurses needed to cover a unit. Rather, the model simply provides a framework within which a Nurse Manager can analyze their needs to set the staffing budget. Staffing requirements are determined by applying a formula that considers each patient's level of care and the length of time it takes to deliver the care.

A staffing budget for a unit is defined in full-time equivalents (FTEs). The Nurse Manager is expected to stay within the staffing budget for the unit on a monthly or quarterly basis, depending on the policy of the healthcare facility.

Staff members in some healthcare institutions have formed a bargaining unit commonly referred to as a union, which negotiates the terms of employment with the healthcare facility for union members. The terms are specified in a collective bargaining agreement, which spells out what management and union employees can and cannot do.

A union member can file a grievance if he or she believes management has violated the collective bargaining agreement. The grievance is resolved through the grievance process where both management and the union present their side of the case.

Quiz

1. What nurse care delivery model specifies that each patient will be cared for by a private nurse?

   (a) Functional model

   (b) Case model

   (c) Team nursing model

   (d) None of the above

2. What is a major objective of the primary nursing model?

    (a) Place the registered nurse back at the patient's bedside

    (b) Place the registered nurse as the team leader

    (c) To recognize that nurses are professional

    (d) None of the above

3. The case management nursing model:

    (a) Uses the team approach to nursing care

    (b) Requires that the nurse case manager provide all the care for the patient

    (c) Moves the patient through the healthcare system in the most cost-efficient manner without sacrificing the quality of patient care

    (d) All of the above

4. The functional nursing model was introduced to address a nursing shortage during World War II.

    (a) True

    (b) False

5. A staffing budget for a unit is defined in full-time equivalents.

    (a) True

    (b) False

6. Staffing requirements for a unit are based on:

    (a) The level of care required by each patient

    (b) The nurse-to-patient ratio

    (c) The judgment of the Nurse Manager

    (d) The judgment of the nurses' union

7. The Nurse Manager can modify a collective bargaining agreement between the nurses union and the healthcare facility.

    (a) True

    (b) False

8. What is a grievance?

    (a) A process used by management to address violations of the collective bargaining agreement

    (b) A complaint that a union member violated the collective bargaining agreement

(c) A complaint that management violated the collective bargaining agreement

   (d) None of the above

9. A Nurse Manager cannot create a past practice.

   (a) True

   (b) False

10. The National Labor Relations Act defines work practices that the Nurse Manager must follow.

   (a) True

   (b) False

CHAPTER 4

Delegation and Supervision

It is the flu season. Your 35-bed medical-surgical unit is full. Two of your five registered nurses (RNs) call in sick an hour before the shift change, and no one from the previous shift is willing to stay overtime. Union rules prevent you from mandating overtime for anyone. The hospital's float pool of nurses is empty, and four hours will go by before the next shift can be called in early. Meds have to be given. Patients have to be prepped for tests. And the flu has left more than a fair share of patients vomiting and with diarrhea. An hour from now you'll have three RNs and two nursing assistants to care for 35 very sick patients. The nursing director's suggestion—deal with it.

As a Nurse Manager, how do you deal with it?

Respond quickly. Develop and implement a strategy whereby you delegate responsibilities to the staff, and then supervise them as they give patient care. Delegating responsibilities can be deceivingly simple, such as assigning patients evenly among the available RNs—the assumptions being that all patients require the same level of care, and that RNs are proficient in every nursing skill. Neither assumption is correct.

Responsibilities are delegated to qualified staff based on your confidence that they are competent to perform those responsibilities with little or no supervision. Other factors also influence how responsibilities are delegated, such as hospital policy, union rules, labor laws, and staff morale. Keep in mind that there is strong evidence that delegating responsibility increases productivity and improves employee morale. A more in-depth analysis of delegation is discussed later in this chapter.

In this chapter, you'll learn how to delegate and how to supervise a staff under normal circumstances and in less-than-ideal situations.

Crunch Time—The Pressure to Do More with Fewer Staff

Nurse Managers are asked to do more with fewer staff because of two overwhelming factors: the economic crises in the healthcare industry, and a shortage of nurses. A continuing rise in healthcare costs is forcing the federal government to come up with alternative reimbursement plans, and employers are pressuring their medical insurance carriers to ease the spiraling expense of health insurance premiums. As a result, changes such as the following are occurring in the healthcare delivery system:

- The Prospective Payment System (PPS) introduced by Medicare in 1983 reimburses hospitals at a fixed rate based on the Medicare patient's Diagnosis-Related Group (DRG), regardless of the hospital's actual expenses. The exception is for Medicare patients who require an exceptionally high cost of care.

- Managed care companies set terms for both financial reimbursement and healthcare delivery by giving patients a financial incentive to use a selected group of healthcare providers. Healthcare providers must agree to reimbursement terms and regulations for providing care in order to join the managed care group. Administrative costs to process claims and the delay in receiving payment typically increase the healthcare provider's expenses.

- The cost of medical insurance is beyond the reach of millions of patients who are without the financial resources to pay for healthcare. Government programs, charities, and healthcare facilities are expected to absorb the cost of their medical care.

- Competition to attract patients and physicians has pressured healthcare facilities to purchase expensive state-of-the-art technology that is likely to be underutilized and costly to operate and maintain.

These and other changes to the healthcare delivery system reduce income and force hospitals to trim costs; consequently, the Nurse Manager has fewer resources available to care for patients.

Compounding the financial squeeze is the shortage of nurses and Nurse Managers. The demand for nurses and Nurse Managers is greater than the supply, and conditions are expected to worsen as baby boomers begin to acquire aged-related conditions. Also, the average age of nurses today is 43.5 years, so we can expect to see many nurses retire in the coming decades. Nurses with the advanced degrees required to teach nursing school are in even higher demand—so the nursing schools cannot meet the demand. Competition to get new graduate nurses is extremely high.

Nurse Managers find themselves responsible for overseeing multiple patient care units and supervising a staff consisting of unlicensed assistive personnel (UAP), temporary RNs and licensed practical nurses (LPNs), and staff RNs whose patient load is stretched to the limit. The only way the Nurse Manager can ensure quality patient care under these conditions is to delegate some of their responsibilities to those they supervise.

What Is Delegation?

Delegation is getting work done through others by transferring responsibility for an activity to another without transferring accountability for the activity. It is important to differentiate between responsibility and accountability. *Responsibility* means you perform the activity. *Accountability* means you are answerable to an authority for the activity, regardless of whether you perform the activity or delegate the activity.

If you are a Nurse Manager, you are both accountable and responsible for the care of all patients in your care unit. You delegate your responsibility to care for a particular patient to a staff RN. This means that the staff RN, who is licensed to practice under the standards of the state's Nurse Practice Act, becomes responsible for the patient's care. Although you no longer are responsible for care of the patient, you still are accountable for the care that is given.

For example, suppose the staff nurse fails to consult the coverage schedule and gives too much insulin to a diabetic patient. The Nurse Manager can be held accountable for the error if they assigned the task to a nurse who had never been educated to read and understand how insulin orders were written in that hospital, or a nurse who had never administered insulin. The Nurse Manager was not responsible for administering the insulin, but is responsible for knowing the skills of the staff and assigning patient care activities to the appropriate staff member. Every RN and LPN is responsible and accountable for his or her own practice.

A key role of a Nurse Manager is to delegate responsibilities to a qualified and competent person. It is critical to realize that a qualified person may not necessarily be competent to take over the delegated responsibility. This is particularly true when activities are delegated to a licensed nurse. The license qualifies the nurse to perform all activities specified in the Standards of Nursing Practice and in specific state Nursing Practice Acts. In reality, the nurse may not have performed the activity in years, and requires revaluation before being competent to perform the activity with the patient. This really becomes an issue when nurses are reassigned from their regular unit to work in a unit with different skill-level requirements. For example, a medical-surgical nurse may be reassigned to labor and delivery.

Sometimes an activity, or a portion of the responsibility for an activity, is delegated twice. First, the Nurse Manager delegates the activity to an RN. The RN then delegates a portion of that responsibility to an LPN or UAP. The RN is still responsible for the activity of the UAP, even though the UAP performs the activity. For example, an RN who is competent to care for patients normally treated on the unit typically delegates evening care to a UAP. The RN—and by extension the Nurse Manager—is accountable if the patient falls out of bed while the UAP performs evening care. The fall will be considered accidental if the UAP has received training and education in total evening care and the appropriate safety measures, such as side rails being in place, were followed. In terms of personal liability, nurses are held to the standard of care of the reasonably prudent nurse employed in the role in question. That means that if you are a "reasonably prudent" nurse, competent to care for the patient population on the unit, and aware of the skills of the UAP, you have delegated appropriately. The outcome for the patient, a fall, will not be considered a shortcoming of your practice.

There are three activities that center on professional judgment that the RN cannot delegate:

- Assessing the patient.
- Formulating the nursing diagnosis and planning patient care. Based on the patient assessment, the RN must determine the nursing diagnosis, set goals for the patient's care, and develop a care plan.
- Performing interventions that require professional knowledge and special skills to perform.

Legal liability transfers with transfer of the responsibility. When a Nurse Manager delegates primary care for a patient to an RN, the RN is legally obligated to give the patient an acceptable level of care. The Nurse Manager must determine that it is reasonable and prudent to delegate a particular task to the RN based on the Nurse Manager's knowledge of the staff nurse, the condition of the patient, and the general state of the unit. Otherwise, the Nurse Manager could also be held liable for the actions of the RN.

It is, of course, expected that an RN will not accept an assignment that they are not educated or are not competent to perform. For example, a psychiatric nurse who is floated to a pediatric unit must refuse to administer intravenous fluids to a pediatric patient. The psychiatric nurse does not have the education or skills to perform this procedure and therefore must refuse the assignment. Every state Nursing Practice Act states that an RN is ultimately responsible for the care they provide to each patient.

Furthermore, some Nurse Practice Acts may hold the Nurse Manager responsible for care given by anyone other than an RN. Other Nurse Practice Acts may be ambiguous, saying that the Nurse Manager must use "acceptable standards" when delegating a task. Your healthcare facility can provide you with the legal meaning of using "acceptable standards." Chapter 7, "Legal Issues," provides more in-depth coverage of these and similar situations.

TASKS THAT CAN BE DELEGATED

It sometimes is difficult to know what tasks can be delegated, especially when a Nurse Manager is delegating a task to an LPN or a UAP. There isn't a definitive list of what tasks can and cannot be delegated. However, the following serves as a guide for the Nurse Manager:

- **State Nurse Practice Act and state regulations**   A state's Nurse Practice Act provides the legal framework within which a nurse can practice nursing in that state. This act, along with other state regulations, give the Nurse Manager a basis for deciding what tasks should be delegated to staff. Most states have a separate Nurse Practice Act for Licensed Vocational Nurses (LVNs), LPNs, and RNs.

- **Job descriptions**   Consult the formal job descriptions for staff members within your healthcare facility. Job descriptions typically define the kinds of tasks each employee is expected to perform. For example, some LVNs/ LPNs can perform limited I.V. therapy, while others cannot.

- **Policies and procedures**   Become knowledgeable about the policies and procedures of your healthcare facility for delegating tasks. These policies and procedures are usually approved by the facility's legal department to conform to laws, regulations, and judicial rulings.

- **Patient needs**   Be aware of the immediate need of a patient and the competency of your staff. A routine task, such as taking vital signs, might be delegated to a UAP while the RN gives urgent care to another patient.

HOW TO DELEGATE A TASK

How to delegate a task might seem obvious—for example, simply tell someone to perform the task of administering a blood transfusion to the patient. However, doing so assumes that the staff member knows how to perform the task according to hospital policy and procedure. This may be the situation with staff members who have proven their competence, but may not be true for new members to the unit, including those who float from a different unit.

Although all RNs are validated to perform many standard tasks, some may have little if any experience actually performing particular tasks. For example, a nurse who floats from a psychiatric unit to a medical-surgical unit may never have administered a blood transfusion. Therefore, before delegating a task, a Nurse Manager must:

- *Clearly understand the complexity of the task.* Know specifically what needs to be done.

- *Know the time required to complete the task.* Use an average time, rather than the shortest time.

- *Know the skill level required to perform the task.* Avoid equating skills with titles.

- *Verify that the staff member has the skill level to perform the task, and has recent experience performing the task.* An inexperienced staff member might not perform the task properly or within the excepted time period.

- *Determine if the staff member has time to perform the task.* Assigning a task to a staff member who is already overloaded carries the risk that the task will not be successfully completed.

- *Develop a contingency plan if the desired outcome isn't achieved.* What will you do if the task is not completed or doesn't have the expected outcome?

- *Establish a level of supervision required for the task.* How much of your time is required to supervise the staff member while they are performing the task?

- *Set criteria for evaluating the outcome of the task.* How will you and the staff member know if the task is successfully performed?

- *Clearly communicate the task to your staff member.* State the task, duration, time frame within which the task is to be performed, and the complexity of the task.

- *Confirm that the staff member accepts responsibility to perform the task.* Every delegation of a task must be accepted by the staff member, otherwise the task is not delegated.

- *Determine the support and equipment the staff member needs to perform the task.* Make sure you provide them.

AN EXAMPLE CASE REQUIRING DELEGATION

A Nurse Manager is responsible for the care of 19 patients in a 25-bed medical unit. Every patient is assigned to an RN, who will call upon UAPs for necessary assistance because the RN cannot provide every aspect of care to each of his or her assigned patients. The following is a breakdown of the current patient situation:

- A patient with abdominal pain of unknown etiology who has just arrived from the emergency department.

- A patient four days post stroke with minimal right-sided weakness and profound speech impairment. The patient frequently cries in frustration and the recently written physician orders are not yet transcribed.

- A diabetic who is on respiratory isolation until tuberculosis is ruled out, and requires a glucose-monitoring meter reading before insulin is administered. The patient is scheduled to receive insulin before every meal.

- An 81-year-old who has congestive heart failure and a stage two decubitus; the patient has been on the unit for 15 days and requires a dressing change every shift.

- A 31-year-old diagnosed with Crones disease and chronic pain; the patient has been on the unit for two days and receives continuous total parental nutrition via a subclavian line with a dressing that must be changed this shift.

- Fourteen additional patients who require routine updating of care plans, charting of vital signs, medication administration, and personal care.

The Nurse Manager must delegate tasks to the staff. The following discussion explains how this may be done for this example.

The admission from ER must be assigned to an RN to complete the admission physical and psychological assessment, and to review and carry out the initial physician orders and the administration of medications. The nurse will administer the ordered pain medication and note the patient's reaction to it. The UAP is assigned to assist the RN and comfort the patient. Patient assessment is always the responsibility of an RN, as is the transcribing of physician orders in most hospitals. The UAP will assist in transferring the patient to the bed and seeing to the provision of necessary hospital supplies (water pitcher, etc.).

The RN will transcribe the physician orders for the stroke patient, and continue with educating the patient regarding potential recovery from the stroke. The UAP will assist the patient with meals and personal hygiene measures.

The diabetic who is on respiratory isolation is assigned to a UAP who has been educated in respiratory isolation techniques and the use of the glucose-monitoring meter. The UAP knows that the supervising RN will need to know the meter reading before administering the insulin.

The 81-year-old patient who has congestive heart failure is assigned to a UAP for all comfort measures and physical care. The RN supervising the UAP will be responsible for the dressing change, as the condition of the decubitus (exudate, size, color, etc.) must be assessed by a nurse and carefully recorded in the patient's chart.

The 31-year-old patient diagnosed with Crone's disease and chronic pain is assigned to an RN, with assistance as needed from a UAP. The RN will need to spend a good deal of time with this patient, as he has a serious illness and is not known to the staff (meaning he has not been hospitalized on this unit previously). He may well be frightened and require a great deal of reassurance regarding the staff's ability to care for him. Having a chronic disease, he will have in-depth knowledge of the type of care he requires. The RN will change the dressing, noting anything out of the ordinary, see to the administration of the total parental nutrition, and discuss issues of pain control with the patient, assuring him that staff nurses have been educated regarding pain control and the hospital has a Pain Management Team if he needs additional consultation. The RN will document on the chart the patient's response to all of the above interventions, as well as the care plan, if appropriate.

The remaining 14 patients are assigned to RNs, who will be accountable for all aspects of care for the assigned patients and for professional activities. The care will be delivered with support from UAPs, as needed.

What Is Supervision?

Supervision, according to the American Nursing Association (ANA), is the active process of directing, guiding, and influencing the outcome of an individual's performance of an activity or task. For example, the Nurse Manager directs the staff member to administer a blood transfusion to a patient and observes whether or not the patient received the transfusion.

Supervision begins once the task is delegated to a member of the staff. The Nurse Manager then observes and guides the staff member through the task to ensure that the task is performed according to expected standards.

There are two types of supervision:

- **Direct supervision**   Occurs when the Nurse Manager observes and directs the staff member through each step of the task. This may occur when a nurse is performing a procedure for the first time, or when a complex procedure that has serious side effects is performed.

- **Indirect supervision**   Occurs when the Nurse Manager oversees the performance of a task. The task is delegated to a staff member, who is then responsible to manage each step of the task independently of the Nurse Manager. The Nurse Manager ensures that the task begins on time and is successfully completed. The staff member is responsible to report to the Nurse Manager any reason that precludes the successful completion of the task.

There is a tendency for Nurse Managers to provide indirect supervision, because staff members perform many routine tasks that don't require direct supervision. However, when tasks become routine and there is pressure to care for an ever-increasing patient load, it may lead to skipping some steps, such as hand washing between patients, that can jeopardize patient care.

Therefore, it is very important that Nurse Managers develop trust but verify—that is, trust that your staff will perform a task correctly, but also periodically verify that the task has been properly carried out. As a general rule, a Nurse Manager should more closely supervise any task that

- Has potential for serious harm to the patient
- Is complex
- May require steps other than those that are normally followed
- May have an unpredictable outcome
- Is not performed frequently
- The staff member has not performed frequently

EVALUATING YOUR STAFF

Evaluating staff is an important aspect of supervising, because the Nurse Manager's evaluations identify strengths and weaknesses of each staff member. A Nurse Manager can build on those strengths and develop a plan with the staff member to strengthen weaknesses through education and modification of performance.

Hospitals typically call these evaluations *performance appraisals*, and the Joint Commission on Accreditation of Healthcare Organizations (JCAHO) mandates that they must be done for each employee on a regular basis. For most healthcare organizations this means once a year at the minimum. Employees' performances are evaluated against standards, and measurements against standards are imprecise. For example, how does the Nurse Manager differentiate between "excellent" and "very good," or "always" and "almost always"?

Performance standards are derived from job descriptions and vary depending on the specialty area of employment. They should be as objective as possible, as clearly defined as possible, and measurable through observation. For example, in the nursery, the standard may be that all babies receive a documented Apgar score within 5 minutes of delivery. In the operating room, it may be that patients receive an antibiotic within 15 minutes of the beginning of an operation. On a medical unit, it may be that care plans for the patient are written immediately after the admission assessment. If the Nurse Manager notes consistent failure to perform these duties, the performance appraisal will highlight the poor performance.

Nurse Managers use anecdotal records during the year to document the staff members' performance, because one cannot remember details regarding behavior when it comes to numerous staff persons. A Nurse Manager should not let the most recent occurrence, whether good or bad, color the appraisal. Instead, they should review the entire year before completing performance appraisals. That is why it is good to do quarterly reviews, whether all the details are shared with the employee or not. Staff members also need to know periodically whether they are doing well, or if there are areas in which they need improvement. A staff member should never be told for the first time at their annual evaluation that they were giving substandard care.

One goal of a performance appraisal is to correct and improve behaviors and skills. This is usually accomplished through remedial education. Another goal is to provide motivation to the employee to improve in areas of weakness. In some hospitals, the appraisals are used to decide salary increases or as evaluations for promotion. Something positive should always be included in the evaluation. Most people are not motivated by negativity. If they know you believe in their ability to do well, most will strive to prove you right.

Evaluations are constant and ongoing, happening in a very active and stressful environment that must balance the needs of patients, the policies of the healthcare facility, and the needs of staff. Staff members are likely drawn from diverse cultures and demonstrate a variety of personalities. Conflicts among competing factors will arise and must be addressed fairly.

MAINTAINING A SAFE ENVIRONMENT

Staff safety is the uppermost concern, because no one on a unit is safe if staff members are not safe. Cleanliness and access to sinks for hand washing will protect staff members and patients from the spread of infection, a top priority in hospitals today. All equipment must be in good repair and working order to prevent accidents.

Walking rounds by the Nurse Manager with someone from the engineering department go a long way toward keeping the environment safe. The bioengineering department must also be alert to breakdowns to prevent injuries to staff or patients from complex electronic equipment. Violence in the workplace is another factor that must be kept in mind. Potential for violence is possible among or between patients, family members or visitors, and staff members. It is an infrequent occurrence, but hospitals do sometimes admit patients who are potentially dangerous to staff, and everyone should carefully monitor the moods and behaviors of patients and their family members as a safety measure. A Nurse Manager should be alert to staff conflicts that can escalate into violence, and consider some continuing education related to conflict resolution as an important investment. Violence should never be tolerated. If someone has a propensity to be violent with a co-worker, they could also be violent with a patient. This is seen more commonly with patients who cannot defend themselves or cannot complain. If a case of violence against a patient occurs and the Nurse Manager knew beforehand that the employee was capable of violence, the Nurse Manager may be sued for damages.

Depending on where you live, the threat of terrorism or a bioterrorism attack is a fact of life in today's world. A Nurse Manager should participate in hospital-wide disaster drills and increase their own knowledge about the likely pathogens considered potential weapons, such as smallpox, anthrax, and plague.

A Nurse Manager must not neglect employee health when it comes to keeping staff safe. If the hospital or facility has its own employee health department, it will offer yearly physicals, immunizations, and various tests for illnesses that may strike hospital employees. Even today, staff members need to be prevented from contracting childhood illnesses such as measles or mumps. More threatening is the possibility of exposure to drug-resistant tuberculosis or HIV, the virus that causes acquired immunodeficiency syndrome (AIDS). The Nurse Manager must stay current with the practice of universal precautions and insist that the health care organization purchase the necessary protective supplies and equipment, especially the devices that protect against needle sticks. If healthcare workers are not protected and healthy, who will care for the public?

Hand in hand with staff safety goes patient protection and safety. The very nature of being hospitalized puts patients at risk for all of the previously mentioned diseases, and organizations must provide a safe, clean environment for patients at all times and in all parts of the hospitalization process. Contagious patients are assigned to isolation rooms to protect others, and all the precautions previously mentioned apply for their safety as well. Of course, one of the biggest risk factors for patients is medical error. Medical errors may occur during invasive procedures in the operating room, as a result of medication error, and, all too frequently, from hospital-acquired (*nosocomial*) infections. It is estimated that 50,000 to 90,000 patients die each year in hospitals due to such errors, and all healthcare workers are potentially culpable in these errors. Many untoward outcomes and even deaths are preventable, and Nurse Managers should work to ensure that everyone on their unit consistently practices the policies and procedures that support prevention. Proper hand-washing by staff is the single most effective measure that can prevent nosocomial infections.

Physicians' orders can sometimes lead to conflict and even to errors. Nurses must carefully transcribe and follow the written physician orders. Mistakes can be made related to handwriting, drugs with similar names, dosages, and abbreviations. Causes of medication error are complex and usually involve poorly designed systems (how the order gets to the pharmacy, how it is labeled, how it gets to the unit, etc.) and mistakes by individuals. Patients with identical last names compound the situation.

Obviously, everyone involved in medication administration needs to keep patient safety a priority in all situations. In rare instances, the physician may write an order that is inappropriate, in which case it is up to the pharmacist or nurse to call this to the physician's attention and have the order corrected. A nurse is autonomous in their own practice, so a nurse may refuse to carry out any order that he or she believes may be harmful to the patient. Staff nurses should always have someone to whom they can turn to receive guidance in such cases. The Nurse Manager can review the situation and intervene and support the nurse in any way that seems appropriate. Every shift has a supervisor to call upon when guidance is required. Nurses still make their own decisions and are legally accountable for their actions. Omissions, or the failure to provide something to the patient, are as egregious as commissions, doing the wrong thing.

Healthcare facility policies impact nursing practice in myriad ways, and thus a Nurse Manager needs to be well aware of all of them. Every unit in the hospital will have a policy manual so that questions may be answered quickly. Does the policy in your hospital permit RNs to start an intravenous line? May nurses give medication "I.V. push?" Are LPNs permitted to start a blood transfusion? The policies for these and many other activities may vary from organization to organization, and perhaps from state to state. Know the policies in your hospital. The absence of a particular policy can cause conflict, but following policy will not. Situations may arise where there is no current policy to provide guidance. In these situations, the nurse, the Nurse Manager, or a supervisor must seek out others in the organization to brainstorm and come to some consensus on a solution.

Hospitals have chief medical directors, ethics committees, and risk management departments to assist everyone. If an unusual and infrequently occurring situation arises in the middle of the night, and it is an issue that cannot be put off until morning, the appropriate experts must be called regardless of the time. For example, suppose the patient's family wants the patient transferred to another hospital at 3 A.M., and the medical staff deems it too dangerous to transfer at the present moment. The charge nurse, the supervisor, and the physician will attempt to dissuade the family, but they may need backup from a patient representative or the assistance of a clergyman. They may need the chief of staff physician to talk with the family. Hospital policy cannot provide guidance for each unique situation, but there will always be other team members to help.

Be cognizant of the personal needs of staff members. Healthcare organizations are generally bureaucratic and hierarchal in nature. The organization operates by policy, which is necessary when hundreds, if not thousands, of employees are involved. A Nurse Manager is guided by written policies and past practice decisions, but nevertheless must be sensitive to unexpected situations and able to deal with them. Staff conflicts arise regarding assignments, schedules, holiday time off, etc. A key leadership skill to develop to resolve these issues is *conflict management*. The Nurse Manager will be able to diffuse conflict by understanding each person's perspective and negotiating to find an acceptable resolution. It isn't always easy, but it is always possible.

When assessing a staff member, a Nurse Manager must focus on the staff member's behavior and not on their personality, because the objective is for the staff member to care for the patient. For example, suppose that an RN remains quiet and keeps to herself throughout the shift. This is atypical of other nurses on staff. However, the RN provides excellent patient care. The Nurse Manager's assessment should focus only on the RN's patient care and not her interpersonal skills with the other staff, as long as the RN's personality is not impeding patient care.

The following are some other ways that a Nurse Manager can cultivate good relations with staff:

- Develop a trusting relationship with your staff that encourages them to share their patient care problems with you. A problem always seems to diminish in size when shared with another person. Avoid creating a work environment where a staff member feels he or she will be penalized for sharing a problem or admitting to an error. The attitude should be to solve a problem or remedy an error together.

- Treat all employees the same way. Avoid any appearance of favoritism because that will cause a division among your staff that is difficult to repair. Hold everyone to the same standard by keeping everyone focused on patient care. Part of becoming a Nurse Manager is realizing that what was appropriate as a staff nurse may not be appropriate as a Nurse Manager.

- Set clear, fair objectives for each employee so they are able to gauge their own performance. The employee will be aware of a performance problem long before you officially discuss their performance.

- Use the performance appraisal system in your healthcare facility to praise behavior that meets the unit's patient care expectations, and to discuss how to improve behavior that falls short of those expectations.

- Don't minimize behavior traits that lead to serious problems with patient care or that jeopardize the working relationship with the staff. Staff members must be team players and be willing to help each other care for patients.

If a staff member does not fit with the team, then the Nurse Manager is responsible for discussing the problem, privately, with the staff member. Perhaps the person is unaware of the behavior that caused the misfit. The Nurse Manager should formulate a plan of remediation with the staff person's input, and allow the staff member time to change the undesirable behavior and to "fit in." If the staff member is unwilling or unable to change, then the Nurse Manager should proceed to discuss finding another position, on another unit perhaps, where the staff member will be more comfortable. For example, a person who is comfortable working on a hospice floor may not "fit" on a critical care cardiology unit. That is the great thing about nursing—there are niches for all types of people.

A Nurse Manager may also consider, with the employee's consent and presence, having an open staff meeting in an attempt to analyze just what is happening among the team members, and formulate plans to remedy the situation.

The reassignment to another unit of a staff member is difficult for both the Nurse Manager and the employee. However, in most situations, the problem can be resolved amicably by considering the employee's situation carefully. If an employee doesn't fit well with a team, they probably realize it and wish there were a way to move to a more compatible team. The employee likely also realizes the reality of changing jobs—lost wages, poor recommendations, and gaps in a resume that raise questions by potential employers.

Here's how a Nurse Manager could handle conflict when it primarily relates to patient care:

- Have a congenial conversation with the employee where you present the performance problem objectively. Compare the goal of several tasks that were delegated to the employee with outcomes. It should become obvious that the employee's performance was inadequate and jeopardized patient care.

- State that the problem lies with the competency in the performance of certain skills and is not related to the intrinsic goodness of employee. This approach helps save face and focuses the employee's attention on finding a solution. The solution usually lies in additional clinical education for the employee.

- Explore other areas within the healthcare industry that you feel would offer a better fit. Don't imply that you or HR has another position to offer the employee. The impetus is for the employee to find a new position for themselves.

- Explain the healthcare facility's policies that apply to this situation. Typically, there are one or more formal negative assessments that go into the employee's file before the employee is terminated. Suggest that you will hold off filing the formal assessment for a few weeks so that the employee has time to find another position within or outside of the healthcare facility.

The meeting should end with the employee focused on finding a new position and understanding that you will be supportive in such endeavors.

SHARING SUPERVISION

The Nurse Manager in many healthcare facilities shares supervision of the staff with other nursing supervisors who take over the Nurse Manager's responsibilities when they are off the unit. The role of a nursing supervisor is broadly defined by some healthcare facilities as supervision of patient healthcare and the administration of the unit for a shift as directed by the Nurse Manager.

It is important that nursing management establish supervisory boundaries for a nursing supervisor that specifically define the scope of the supervisor's responsibility. This simply defines what the nursing supervisor can and cannot do in their supervisory role.

Generally, the nurse supervisor can do the following:

- Delegate responsibilities to the staff

- Set schedules during the shift such as breaks and mealtime

- Request replacement staff if a staff member is absent from the shift

- Provide the appropriate level of staff supervision

- Interact with other departments of the healthcare facility to ensure proper patient care

- Correct employee behavior
- Relieve a staff member from responsibilities for the shift for cause
- Provide the Nurse Manager with input for an employee's evaluation

In most cases, the nurse supervisor cannot do the following:

- Terminate a staff member
- Formally evaluate an employee
- Address issues regarding personnel matters such as compensation, benefits, advancement, termination, and other activities not directly related to patient care
- Give directions that counter directions given by the Nurse Manager

The Nurse Manager delegates certain management responsibilities to a nurse supervisor to enable them to direct the staff in giving patient care while the Nurse Manager is away from the unit. While a Nurse Manager is in charge of one patient care unit, the evening and night supervisors are in charge, for the duration of their shifts, of several units in a given division. A small hospital may have only one off-shift supervisor on duty who supervises many nurses, primarily because the Nurse Manager is not on the unit. The Nurse Manager does not directly supervise the supervisor. The relationship is one of collegiality and each works to support the other.

The Nurse Manager relies on feedback from the nurse supervisor regarding the performance of the staff; and the nurse supervisor relies on the Nurse Manager to assign the appropriate number of staff for the shift, ensure that they are educated and trained for the work of the unit, as well as making sure the entire unit environment is ready for the evening and or night shift work. Supervisors solve on-the-spot problems, but may call the Nurse Manager at home if necessary. A Nurse Manager who is not comfortable with conflict must not use the supervisor to do the "dirty work." It is unfair to everyone.

The Nurse Manager shares the unit goals and objectives with the supervisors and enlists their support in meeting them on the evening and night shifts. Conflicts do arise and should be discussed in face-to-face meetings in order to come to a joint resolution and to prevent similar conflicts going forward. For example, to cover a sick call on the night shift, the supervisor may ask an evening shift nurse to work the night shift that follows. The supervisor then promises that the nurse can be off the next evening. When this happens, the Nurse Manager must spend some time during the day finding coverage for the evening nurse who was granted the evening off. The Nurse Manager may be unhappy, but should understand the dilemma that faced the supervisor. If the Nurse Manager thinks there was a better alternative, they should suggest that alternative for when the situation arises again. A Nurse Manager should avoid becoming a "Monday morning quarterback" to the person in the thick of the action (the supervisor) when they make a decision. It may come to pass that, when all the data is gathered, a different decision would have been more appropriate, but only in rare instances would a decision need to be overturned.

Conflicts may occur when the nursing supervisor makes a decision affecting employees that is counter to a decision that the Nurse Manager would have reached under similar circumstances. The Nurse Manager must give a nurse supervisor the latitude to make a decision without second-guessing it, as long as the decision is reasonable and within the scope of the nurse supervisor's responsibility. The Nurse Manager should voice concerns about the decision privately with the nurse supervisor and support the decision publicly. Any lack of public support will undermine the nurse supervisor's ability to manage the staff.

A Nurse Manager should plan to work as a team with their nurse supervisors and share information that helps them care for patients during the shift. Nurse Managers should maintain an open and candid relationship with nurse supervisors. This helps to foster team spirit and encourages working toward the same goal.

Summary

Nurse Managers are asked to do more with fewer staff because of two overwhelming factors: the economic crises in the healthcare industry, and a shortage of nurses. Therefore, the Nurse Manager must delegate tasks to nurse supervisors and staff members, and then supervise their activities to ensure that proper care is given to patients.

Delegation means to get work done through others by transferring responsibility for an activity to another person without transferring accountability for the activity. The Nurse Manager is answerable to a more senior nursing administrator for each activity, whether performed personally or delegated to another.

There are three activities that center on professional judgment that RNs cannot delegate: patient assessments, formulating the nursing diagnosis and planning patient care, and performing interventions that require professional knowledge and special skills.

A Nurse Manager should consult the state's Nurse Practice Act, state regulations, and the policies and procedures of their healthcare facility to determine what tasks can be delegated and to whom. Legal liability transfers with transfer of the responsibility. The state's Nurse Practice Act may hold the Nurse Manager responsible for having delegated care to be given by anyone other than an RN.

Supervision is the active process of directing, guiding, and influencing the outcome of an individual's performance of an activity or task. Supervision begins once the task is delegated to a member of the staff. There are two types of supervision: direct and indirect. Unless a task may jeopardize patient care, the Nurse Manager should provide indirect supervision of the staff.

The Nurse Manager must set goals for the staff, determine how to measure those goals, and assess the staff's performance on a regular basis. Assessment identifies each staff member's strengths and weaknesses and then details a plan for strengthening the weaknesses.

Quiz

1. A new patient arrives on the unit. Which staff member is delegated the responsibility to assess the patient?

   (a) Nurse supervisor

   (b) Registered nurse

   (c) Licensed practical nurse

   (d) Nursing assistant

2. What governs tasks that a Nurse Manager can delegate to a nursing assistant?

   (a) Nurse Practice Act

   (b) Policies and procedures of the healthcare facility

   (c) Training and competency of the nursing assistant

   (d) All of the above

3. Before delegating a task, the Nurse Manager must:

   (a) Have a clear understanding of the complexity of the task

   (b) Know the skill level required to perform the task

   (c) Verify that the staff member has the skill level to perform the task

   (d) All of the above

4. Telling a staff member how to perform a task is an example of indirect supervision.

   (a) True

   (b) False

5. The Nurse Manager should adopt a trust but verify approach when supervising the staff.

   (a) True

   (b) False

6. When a Nurse Manager disagrees with a nurse supervisor, the Nurse Manager should:

   (a) Assess whether the nurse supervisor's decision is reasonable and within the scope of his or her role

   (b) Admonish the nurse supervisor

   (c) Tell the staff to ignore the nurse supervisor's decision

   (d) Override the nurse supervisor's decision

7. A nurse supervisor has the right to relieve a staff member from patient care if the staff member places the patient at risk.

   (a) True

   (b) False

8. The Nurse Manager is:

   (a) No longer legally responsible for patient care once patient care is delegated to a staff member

   (b) Legally responsible if responsibility for patient care is delegated to an incompetent staff member

   (c) Should not be in contact with the patient after the patient's care is delegated to a staff member

   (d) Authorized to overrule policies and procedures of the healthcare facility

9. A Nurse Manager should determine the support and equipment the staff member needs to perform the task before delegating the task.

   (a) True

   (b) False

10. The Nurse Manager should provide direct supervision for complex tasks.

    (a) True

    (b) False

CHAPTER 5

Effective Communication and Conflict Resolution

A national director of nursing for a multi-state healthcare company experienced what some might call a moment of miscommunication. She e-mailed a regional directive to her regional directors of nursing to attend a 9 A.M. teleconference to discuss the pending management reorganization. Nearly half of the regional directors failed to attend. She issued a taut e-mail warning. Her message was clear: Miss another meeting and face disciplinary action. Regional directors were puzzled by her second e-mail. They were present for the meeting. It was the national director of nursing who didn't attend.

The confusion stemmed from the invitation. Everyone assumed 9 A.M. meant their time zone. It didn't. The national director is based in California and many local directors are based on the East Coast. When the East Coast local directors dialed into the teleconference at 9 A.M. their time, the national director wasn't at the meeting—it was 6 A.M. in California. When the national director dialed into the teleconference at 9 A.M. pacific time, the East Coast local directors were at lunch— it was noon.

The national director placed herself in a very unsettling position. She couldn't simply say to the regional directors "never mind," because she had already accused, tried, convicted, and sentenced them for insubordination. An apology could not undo the ill feelings that were conveyed in her second e-mail. Two problems existed. First, the directive ineffectively communicated the meeting time. Compounding this problem, no attempt was used to resolve the conflict after some staff members were not present for the meeting.

In this chapter, you'll learn how to avoid these and other communication problems by learning how to effectively communicate with your staff, colleagues, and physicians, as well as patients and their families. You will also learn proven techniques for resolving conflicts that are bound to arise when you accept the role of a Nurse Manager.

The Art of Communication

For purposes of this discussion, *communication* is the transmission of information from a sender to a receiver using shared symbols to change the receiver's behavior or to convey information. A symbol is a word, a phrase, body language, or other means used to communicate with another person.

For example, a physician writes an order for medication that is read by the nurse, who then prepares the medication and administers it to the patient. The physician is the sender and the nurse is the receiver. The medication order is information. Preparing and administering the medication is the behavior that occurs as a result of the communication. Good communication is two-way; for example, if the physician's orders are not clear, the nurse communicates with the physician to get clarification.

Information is communicated by using shared symbols that are understood by both the sender and the receiver. Some symbols nearly everyone recognizes, such as "call your physician if your temperature rises above 100°F." Other symbols are meaningful to some people and not to others, such as a physician telling an ER nurse to give an alcoholic patient a banana bag. The patient's family may not know what this is, but the ER nurse knows that a banana bag is an I.V. of vitamins, thiamine, and dextrose.

It is critical that the sender use symbols that are recognized by the receiver; otherwise, the information being conveyed might be misunderstood. For example, the physician should realize that a nurse new to the ER may not realize that a banana bag is a type of I.V.

METHODS OF COMMUNICATION

Communication is sent through speech and nonverbal behavior. When the sender is speaking, we listen to their words and observe their behavior to fully understand the information that is being expressed. This happens when you examine a patient. For example, as you zigzag the stethoscope down the patient's back, you may backtrack for a moment for a second listen to lung sounds. You tell the patient that their lungs sound normal, but your behavior raises doubt in the patient's mind. "If my lung sounds are normal, then why did you backtrack? You must have heard something unusual that you are not acknowledging." The patient may think you heard a sign of something serious that you want to confirm before telling the patient.

Be aware of how your words and behavior might be interpreted by the receiver and immediately take steps to clarify any possible misinterpretation. If your behavior contradicts what you are saying in even the slightest way, then you need to explain your behavior. For example, in the scenario above saying something like, "Your lungs sound healthy. Have you been exercising?" implies a positive reason why you might have listened a second time. We, as healthcare providers, must understand that patients are anxious, and should do our best to put them at ease.

Nonverbal behavior can distract the receiver from listening to what you are saying and therefore disrupt effective communication. Suppose that as a Nurse Manager you call an impromptu closed-door meeting with one of your nurses. Your nonverbal behavior sends a signal that could be interpreted as impending trouble for the nurse. The nurse is likely to be on edge when the meeting begins and consequently may pay little attention to what you are saying.

Anticipate how your nonverbal behavior might disrupt effective communication and then take steps to address those potential disruptions before they occur. For example, you might say, "Are you available at 10 A.M.? There are some training opportunities I want to discuss with you." This places the nurse at ease and in the right mind-set to listen to what you have to say rather than preparing to react to criticism.

The tone of your voice adds meaning to your words. Therefore, you should use a tone that complements the information you are communicating. A strong tone is used to give directions, such as giving instructions during an emergency. Hesitation implies that you are unsure and inviting suggestions. You might hesitate when discussing with a colleague whether a new nurse should be assigned to a very sick patient.

The colleague is likely to give an opinion. A carefree tone suggests a less serious situation and could be used to arrange a meeting to discuss training opportunities with your staff. Using an inappropriate tone can lead to miscommunication. For example, if you hesitate while giving instructions in an emergency, the staff is likely to second guess you rather than follow your direction.

Be aware of the timing of when you deliver your message. For example, scheduling a closed-door meeting with one of your staff at the end of shift at the end of a pay period can send the wrong message and raise unwarranted fears of termination. Never tell the nurse at 7 A.M., "Come to my office before lunch. I need to talk to you," if you want that nurse to be at their best on the job that morning.

WRITTEN COMMUNICATION

Written communication requires the receiver to interpret your words without the opportunity for clarification or the ability to hear your tone of voice, which presents a distinct disadvantage over spoken communication.

When you say something, you wait for specific acknowledgement from the receiver so that you know that the information was successfully communicated. If the receiver behaves unexpectedly, you realize that the information was not successfully communicated and can proceed to clarify your statement.

You don't have this immediate feedback when you send your message in writing. It is therefore critical that you carefully read any written communication before you send it to determine if your message might be read differently from what you intended (as in the case of the national director of nursing arranging for a 9 A.M. meeting without specifying the time zone). It is also good to have someone you trust read your correspondence before you send it. A good rule is to use someone of equal or higher rank in the organization for this purpose.

Make sure that you use words that are shared symbols with the recipient. Otherwise, the recipient will have to guess at what you mean because there is no immediate opportunity to ask for clarification.

If you receive an unexpected response to your message, assume that your message has been misread. Give the person a call so you have the opportunity to clarify any misunderstandings. Avoid the tendency, especially with instant messaging and email, to respond in writing quickly.

Keep written communication short and to the point. State the purpose of the document in the first sentence, followed by the facts that you want to communicate to the receiver. Include sentences that clarify any possible misreading of the facts. End with a sentence that tells the receiver how you expect them to behave, such as "Please confirm our meeting." It is also a good idea to send an e-mail after a face-to-face meeting to restate what was talked about and what was decided. If the meeting is between a boss and a subordinate, the subordinate may be nervous and unable to concentrate on everything being said. A written confirmation helps to reinforce what was discussed.

CHOOSING THE BEST MODE OF COMMUNICATION

Should you deliver your message personally or in writing? Your choice depends on a number of factors. Information that is detailed, important, or personal should be in writing, because it can then be referenced in the future. This includes policies, procedures, personnel records, and other legal or quasi-legal documents. Information that is less complex or informal can be communicated verbally, such as simple instructions.

Sometimes both written and verbal communication is appropriate, as in the case of a salary increase or disciplinary action. In these situations, verbal communication is used to introduce and informally explain the written communication to the receiver. The written document is the formal and legal communication. For example, a Nurse Manager might say, "I tried hard to get you the maximum salary increase" when handing a nurse the written notification of the pay increase. Official notification is made in writing.

Deciding between written and verbal communication is also determined by what you want to achieve and what the receiver needs to know. If a Nurse Manager wants a nurse to respond quickly to a situation, then the Nurse Manager should tell the nurse what they want done and provide information the nurse needs to carry out the task. This occurs in many healthcare facilities, where in an emergency, a nurse is permitted to act on a verbal medical order from a physician before the order is written into the patient's chart.

BARRIERS TO EFFECTIVE COMMUNICATION

Several factors may interfere with effectively communicating your message to the receiver:

- **Lack of shared symbols**   You may use vocabulary in your message that the receiver doesn't understand.

- **Distribution**   Your message may be passed through too many hands before the message is received. This is commonly found in large organizations where directives flow through a chain of command that could delay transmission or modify the communication.

- **Clash of personalities** A clash of personalities between you and the receiver can taint how the message is received. Some receivers may have goals counter to your own. A power struggle might exist where receivers hoard information. Be sensitive to the various communication styles. It is not the job of the employee to adjust to a Nurse Manager's personality. It is, however, the Nurse Manager's job to be sensitive and to try to communicate with all personalities effectively.

- **Lack of skills** A lack of basic skills, such as reading, writing, and listening, can silently block communication, since the receiver might go to great lengths to hide these weaknesses. This is especially true of computer and keyboard skills. A receiver may be slow to respond to your email simply because they are uncomfortable using a computer or cannot type.

- **Defensiveness** Avoid situations that anger the receiver or place the receiver on the defensive, because the receiver will focus on self-protection rather than understanding your message.

- **Bad timing** Deliver your message when the receiver is most receptive to listening to you. You would not try to teach your patient how to give their own insulin injections after a long session with physical therapy. A Nurse Manager should not try to have effective communication with employees at the end of their shift. They are tired and want to go home. Also, a Nurse Manager may consider paying them for coming in outside their schedule, which communicates that the Nurse Manager values their time.

COMMUNICATING EFFECTIVELY

An effective communicator clearly states the main point at the beginning of the message, follows up with secondary points, and then questions the receiver to make sure that the message is understood.

Wait for the receiver to digest and react to the message. Be open to questions, because this gives the receiver a way to explore your message and provide you with new information. Take time to understand the receiver's viewpoint. Don't jump to conclusions. Keep the discussion on the topic of the message.

Summarize your message and any information provided by the receiver. Acknowledge areas of agreement and disagreement, and restate questions that remain unanswered. Define the next step in the process.

LISTENING EFFECTIVELY

A good communicator is also a good listener, because twice as much information is gained by listening than by talking. Effective listening occurs when you focus solely on the speaker. Establish eye contact and assume body language that shows interest in what is being said.

Focus on what can be learned from the speaker's message, rather than on the speaker. Sometimes the message is important but the delivery is weak. Listen intently. Avoid reading phone messages, email, or other distractions when listening to someone.

Evaluate each fact contained in the speaker's message. Prepare probing questions that elicit other useful information from the speaker. Give constructive feedback that explores the message. Avoid placing the speaker in an embarrassing situation. Always give the speaker a way to save face.

DISTORTED COMMUNICATION

Distorted communication occurs when your message is not effectively received or is disrupted. Distortion happens in a number of ways, such as delivering your message when there are too many distractions. Your message can also be distorted if the message is delivered too quickly, leaving no time for the listener to digest the message or ask for clarification.

Messages can miss their mark if the sender is difficult to understand because of an accent. This occurs frequently in situations in which the speaker isn't aware that their accent is impeding effective communication. Many times the receiver is embarrassed to ask the speaker to repeat the message, especially if the speaker is the Nurse Manager. Once, a student nurse was asked by a Nurse Manager to check a patient's I and O (fluid intake and output), and the student asked, "Do you want me to use an otoscope?" The student thought the request was to check the eyes and nose. The student was new, and did not yet know the terms and initials used on the unit. This was truly a miscommunication by the Nurse Manager, and a great example of what might have occurred if the student hadn't verified the request.

Messages are also distorted when English is either the sender's or the receiver's second language and unfamiliar idioms are used in the message.

Written communication can be distorted when the message is unclear, incomplete, or inaccurate, as illustrated in this memo:

> To All Staff:
> Please be aware that holiday request forms will be distributed Nov. 15th and must be returned by Nov. 10th.
> Thank you,
> Nurse Manager

This memo assumes that the staff knows to use a holiday request form to schedule time off. Furthermore, the memo doesn't say where to return the completed form. And obviously, forms distributed on November 15th cannot be returned by November 10th.

Punctuation can lead to distorted written communication. An excellent resource is a book about punctuation, titled *Eats, Shoots and Leaves: The Zero Tolerance Approach to Punctuation*, which provides great insight into the importance of punctuation.

Be aware that cultural differences might also impact the effectiveness of your communication. In some Asian cultures, nodding and saying yes is a sign of respect and not necessarily an affirmation. Thus, when an Asian patient nods their head and says yes in response to a question, it does not necessarily mean that they agree or even understand the question. It is therefore important for Nurse Managers and staff members to become familiar with the cultural practices of the community served by their healthcare facility so that they can effectively communicate with patients and their families.

NURSE MANAGERS MUST BE GOOD LISTENERS

A Nurse Manager must be an active listener when dealing with staff. When the Nurse Manager listens, the staff feels that the healthcare facility is listening to them and is interested in their concerns. A Nurse Manager must set aside time when they can give the staff member their full attention. Listening with an open posture and a neutral or positive expression, nodding to indicate understanding, are practices of a good listener.

A Nurse Manager should not do the following when listening to staff:

- Cross arms
- Interrupt
- Finish the person's sentence
- Cause a distraction by looking at something on the desk or outside the window or by answering the telephone
- Use excessive hand movements, which can be misunderstood as aggressive

Conflict Resolution

Nursing is fraught with potential conflicts that must be resolved amicably in order to provide quality patient care. For example, a physician may order medication that is not available in the pharmacy. An RN may disagree with their patient assignments for the shift. Family members may try to stop the respiratory therapist from administering a treatment ordered by the physician.

The Nurse Manager is frequently called upon to resolve conflicts. In a perfect world, the Nurse Manager would listen to both sides and then makes a decision that both sides would follow. This rarely works in the real world, because the Nurse Manager may not have the authority to force their decision on either party. Instead, the art of persuasion must be used to find a solution that both parties can voluntarily agree on.

Conflicts occur when two parties disagree in principle or over an issue. A disagreement in principle is based on deep-rooted fundamental values, such as beliefs regarding termination of a pregnancy or administration of blood transfusions. A disagreement in principle is difficult to negotiate, because the disagreement is not linked to the situation at hand. For example, a patient may refuse blood transfusions based on religious beliefs, regardless of the potential benefits to his or her health. Ethical issues may require the services of the hospital's ethics board.

A disagreement over an issue is usually not deeply rooted and relates to the immediate situation. These disagreements are easier to negotiate. For example, to avoid a delay in treating a patient, a Nurse Manager might persuade a physician to order medication that is available from the pharmacy rather than medication that has to be acquired from outside the facility.

THE NURSE MANAGER'S ROLE

The Nurse Manager's role in a conflict is that of a *mediator*. A mediator is a person who remains impartial and focuses on creating an environment in which both sides can resolve their differences. A mediator is a person who defines the process to reach a resolution and helps both parties use the process to resolve the conflict.

Typically, conflicts arise in a setting that is not conducive to settling a disagreement. For example, an RN may have a disagreement with a physician by telephone at the nurse's station. Family members may confront the respiratory therapist at the patient's bedside. A nursing assistant may complain about his or her schedule to a nurse supervisor in the hallway.

If you are a Nurse Manager involved in such a conflict situation, your initial step is to take the parties to a quiet area and point out that they have not chosen a proper place to resolve their differences. Schedule a meeting to discuss the situation with both sides. The meeting should take place soon, but not immediately. Give both sides an hour or so to cool down. Hold the meeting in a neutral place that doesn't favor one side or the other, such as a conference room that is away from interruptions. Arrange for a teleconference if a party is away from the hospital.

In the meeting, keep the discussion focused on the issue and not personalities. For example, explain to the physician that a patient's treatment will be postponed four hours until the pharmacy acquires the medication that the physician ordered, unless the physician orders an equivalent medication that is stocked by the pharmacy, in which case treatment can begin in 15 minutes.

Begin the discussion by reviewing facts that are not disputed, such as briefly reviewing the patient's condition. This serves several purposes. First, this helps define your role as the neutral party in the discussion. Second, it gets the discussion focused on the patient and not the dispute. After each point, ask both sides if what you said is true. No doubt they'll agree. By doing this, you are setting the scope of the dispute and demonstrating—without saying—that only one of many points is in dispute.

Next, let each side explain their position in turn. Discourage emotional outbreaks and interruptions. Listen carefully to each argument, and try to find additional common ground. Clarify each position by restating what you heard. Indicate additional points that both sides seem to agree on. Ask each side if your understanding is correct. Allow time for both sides to offer a correction, and then summarize the status of the disagreement. This narrows the scope of the problem to the point that is in dispute.

Next, ask for possible solutions. Help both sides think through the advantages and disadvantages of each solution. Conduct a reality check. Is the solution realistically in the best interest of the patient? At this stage, the parties usually find an acceptable solution. The solution may not be perfect, and both parties may still disagree, but the solution addresses the issue. For example, the physician may agree to order the medication stocked by the pharmacy for this patient, if the pharmacy agrees to stock the originally ordered medication in the future. You may not have authority to commit another department to an agreement, but you can inform the parties about how to go about making such a request.

STAGES OF ADOPTION

A common pitfall of conflict resolution is to not give the parties sufficient time to adopt a solution. Although a solution might be obvious to you, each party must buy into the solution by working through the stages of adoption.

The first stage of adoption is awareness. The parties may not be aware of a solution, so you'll need to explain the solution in detail. The next stage is exploration. This is when each party takes a superficial look at the solution to determine if the solution is a possible resolution to the conflict.

Once the solution is considered viable, each party moves into the examination stage and takes a detailed look at the solution, trying to uncover problems that may invalidate it. If the details are sound, the party will want to test the solution under various scenarios before finally adopting it. For example, a physician might be unaware of an alternative to the ordered medication until the Nurse Manager mentions it. The physician might respond by asking general questions about the alternative medication, such as how the side effects compare with the ordered medication. If the side effects are comparable, then the physician will probably examine the alternative by asking how it works and how it worked with other patients before agreeing to prescribe the other medication.

It is critical that you give each party time to work through the stages of adoption; otherwise, it will be difficult for them to agree to your solution.

WHEN BOTH SIDES DON'T COME TOGETHER

In the real world, there are times when both parties are too disturbed to discuss the situation in the same room with the Nurse Manager. Under this condition, it is best to meet separately with each party in a private space to listen to their views about the conflict.

A private meeting gives each party the freedom to express their concerns without fear of contradiction from the other party. As a Nurse Manager, you would begin the meeting by letting the party vent. Steer the conversation to the substance of the issue. Become the fact finder. Reconstruct the situation surrounding the issue. List what the party believes are the facts, and then list points of contention. Don't be judgmental at these private meetings. Keep an open mind. Acknowledge the party's concerns, but don't agree or disagree, because you must remain impartial and neutral. You must stress that the patient is the primary focus and concern.

At the end of both private meetings, verify what are believed to be the facts surrounding the issue. Sometimes emotions cloud what people see as the facts. Your role brings a sense of reality by objectively separating fact from misperceptions.

Based on your private meetings, you'll be able to identify common ground and issues of disagreement. Bring both parties together. Restate the facts. Clarify any misperceptions that might have been mentioned during a private meeting without identifying the party who perceived the situation in this way. Identify common ground and areas in disagreement. Restate each party's position on the disagreement. This gives each side the feeling that their concerns have been heard.

At this point both sides await your ruling. Your job is not to resolve the conflict by edict, but to help both sides reach an agreement. Collectively explore possible solutions and help to lead them to a solution that best helps the patient.

Let's say that a nurse is charting nurses' notes into the computer when another nurse interrupts, saying, "I need to get a stat medication order sent to the pharmacy right now, so sign off!" Both nurses enter into a shouting match. The Nurse Manager meets with each nurse privately, giving each the opportunity to present their view. During the meeting, the Nurse Manager discusses what action the nurses feel is in the best interest of the patient. After concluding the private meetings, the Nurse Manager brings together both nurses and restates each nurse's position. This gives each nurse the opportunity to understand the other's point of view. The Nurse Manager then states the facts and identifies the mutual goal—patient care. With cooler heads, both nurses probably will draw the same conclusion from the facts and devise a way to avoid such conflicts in the future.

CHANGE AND CONFLICTS

Change is often the precursor to conflict. Change is typically initiated by hospital administration and implemented by the Nurse Manager. The Nurse Manager may not agree with the decision and may believe the decision is ill conceived; however, the Nurse Manager must manage the change regardless of feelings to the contrary.

The Nurse Manager's staff looks to the Nurse Manager for creditable, current information about how to implement changes. Therefore, the Nurse Manager needs to develop a strategy to address concerns and minimize conflicts that might be brought about by the change.

A good strategy is to explain the need for change to the staff and describe how the change will affect them and the patients. Let's say that the hospital administration decides to close the dialysis center because there are too few patients who require treatment. The Nurse Manager should be prepared to explain to the staff that current patients will be treated by the underutilized dialysis center at a neighboring hospital, and that efforts will be made to either reassign the dialysis center staff within the hospital, or place them with the dialysis center at the neighboring hospital. There may be those who want to sabotage the efforts. The Nurse Manager must be prepared to deal with potential problems swiftly and effectively, while still responding to staff concerns. It's quite a tightrope act, but it is a management skill that can be learned.

The Nurse Manager must try to anticipate the concerns of the staff. For example, they need to know when the dialysis center will close, how reassignments will be implemented, salary and benefits offered by the neighboring hospital, and other factors that affect their livelihood.

MINIMIZING THE OPPORTUNITY FOR CONFLICTS

Conflicts cannot be avoided, but a Nurse Manger can minimize them by keeping lines of communication open with the staff. The following are some suggestions:

- Hold regular staff meetings and one-on-one conversations. Don't forget to include the night shift.
- Encourage staff participation in the decision process where possible, because this gives the staff a feeling of ownership in the decision.
- Persuade the staff to give feedback and propose ideas to better the operation of the unit.

- Announce changes well before implementing them.

- Introduce changes incrementally. The staff can quickly adopt small changes more easily than involved changes. Be truthful about the effects of the change.

- Break any bad news related to the change as soon as possible. This gives both the Nurse Manager and the staff time to deal with the negative aspect of the change.

CRISIS MANAGEMENT

Occasionally, a conflict rises to the point of a crisis that jeopardizes the effective operation of the unit and the quality of patient care. As a Nurse Manager, you would be responsible for the management of such crises. Crisis management starts by having the right attitude. It is the Nurse Manager's job to take control over the situation, stay objective, and avoid becoming emotionally involved. Your calm demeanor and quiet, authoritative speaking voice will positively affect the behaviors of those involved in the situation.

The following are some additional ways that a Nurse Manager can manage a crisis:

- Quickly assess the situation by listening to everyone involved. Trust what is being said, but verify it. Find the facts of the situation.

- Follow the strategies discussed in this chapter for resolving conflict. Remain flexible and be willing to change direction. Don't point fingers. Focus on getting everyone back on track.

- Make sure that you have confidence builders in your plan to resolve the conflict. A *confidence builder* is something you do that garners the confidence of both parties to the conflict. These include treating each party equally and fairly, letting each side explain their position, and identifying the points that are in agreement and those that are not.

- Build on confidence builders to create a positive trend. Have both parties agree to meet with you either privately or together. Identify facts that both parties can agree on. Identify issues that both parties agree to disagree on.

- Don't be afraid to set a deadline. Look for a stop-gap measure that resolves the crisis and ensures quality care for the patient while both sides work toward resolving the conflict.

Summary

Communication is the transmission of information from a sender to a receiver using shared symbols to change the receiver's behavior or to impart information. Information is communicated by using shared symbols that are understood by both the sender and the receiver.

Communication is conveyed through speech and nonverbal behavior. We both listen and observe behavior to fully understand the information that is being expressed. Nonverbal behavior of the speaker can distract the receiver from listening to what is being said and disrupt effective communication.

Written communication requires the receiver to interpret your words without the opportunity for clarification or to perceive your tone of voice. You, as the writer, don't have immediate feedback to determine if the information was successfully communicated to the receiver.

An effective communicator clearly states the main point at the beginning of the message and then follows up with secondary points and waits for the receiver to digest and react to the message. Distorted communication occurs when your message is not effectively received or is disrupted.

Conflict resolution involves the art of persuasion to help parties to a conflict resolve their differences. Conflicts occur when two parties disagree in principle or over an issue. A principle is based on deep-rooted fundamental values, and an issue relates to a particular situation.

The Nurse Manager mediates a conflict by remaining impartial, and creates an environment where parties to the conflict can resolve their differences. The Nurse Manager clarifies each position and indicates points on which the parties are in agreement, and those for which there is still disagreement, and then encourages the parties to devise solutions to the conflict.

Quiz

1. Fact finding is:

    (a) When both parties state the facts related to a conflict

    (b) When the Nurse Manager identifies and verifies facts related to a conflict

    (c) When one party states the facts related to a conflict

    (d) None of the above

2. The objective of conflict resolution is to:

   (a) Resolve all issues related to a conflict

   (b) Provide a judicial forum to resolve a conflict

   (c) Resolve the conflict sufficiently to provide quality patient care

   (d) Resolve issues favoring one side over the other side

3. Which of the following Nurse Manager behaviors is likely to escalate a crisis situation:

   (a) Objectivity

   (b) Emotional involvement

   (c) Calmness

   (d) Modulated speech

4. The Nurse Manager takes on the role of a mediator in a conflict.

   (a) True

   (b) False

5. Effective listening occurs when you focus solely on the speaker.

   (a) True

   (b) False

6. A shared symbol:

   (a) Is a word, a phrase, body language, or other means used to communicate with another person

   (b) Must be understood by the sender and the receiver

   (c) Is a way to convey information

   (d) All of the above

7. Communication is conveyed through speech and nonverbal behavior.

   (a) True

   (b) False

8. Written communication:

   (a) Requires the receiver to interpret your words without the opportunity for clarification

   (b) Has a distinct disadvantage compared to spoken communication

   (c) Doesn't give you immediate feedback

   (d) All of the above

9. Information that is detailed, important, or personal should be in writing.

   (a) True

   (b) False

10. Sometimes using both written and verbal communication is appropriate.

    (a) True

    (b) False

CHAPTER 6

Policy

As a Nurse Manager, policies define practically everything you and your staff can do in the healthcare facility and how it is done. There are policies that specify how diagnostic tests are performed, how to hire employees, and how to administer the medications the pharmacy stocks. Policies are laws that govern operation of the healthcare facility.

Nurse Managers uphold policies and recommend new ones that address problems that staff members face each day while caring for patients. Nurse Managers work rules for the staff to follow in order to comply with the facility's policies. Violation of a policy can result in disciplinary action that may potentially be appealed to senior management (and possibly to the court system). The Nurse Manager must defend enforcement of the policy.

This chapter explores policies in a healthcare facility and how they are created and enforced, how to create work rules, and the politics of policies.

What Is a Policy

A *policy* is a rule of practice that supplements bylaws of a for-profit or not-for-profit corporation. In essence, policies are the day-to-day rules that govern how managers, affiliated healthcare providers such as physicians, and the staff care for patients and operate the healthcare facility.

A policy has legislative roots. State lawmakers enact legislation that authorizes a state agency to grant a group of individuals the right to form a healthcare corporation. In doing so, the group files a legal document, called *articles of incorporation*, that establishes the corporation's purpose and structure, and the state grants the corporation a *charter* that defines, among other things, what the corporation can and cannot do.

The articles of incorporation and the charter contain broad, general goals and objectives for the corporation. Specific rules and regulations on how the corporation conducts business are contained in the corporation's bylaws. A *bylaw* is a basic rule for the conduct of the corporation's business and affairs, established at the time that the company is formed. For example, bylaws specify the number of trustees on the board of trustees and the process for electing trustees to the board. Bylaws also create positions, such as a chairperson of the board of trustees, president, and vice presidents of the corporation, and specify the roles and responsibilities of those positions. At least one bylaw authorizes the board of trustees to make policies.

Creating a Policy

A policy is created according to procedures described in the bylaws of the corporation. Typically, a policy evolves from new laws or rulings by the courts. Sometimes a policy stems from improvements in operation suggested by the staff, or demonstrated in other healthcare facilities.

The policy is a written document that is usually proposed by a trustee and approved by the majority of the board of trustees. Policies are also created in response to regulatory mandates from entities such as the Joint Commission on Accreditation of Healthcare Organizations (JCAHO), or in response to changes internal to the organization. An example of the latter would be the opening of a cardiac surgery program requiring written policies on the role of physician assistants in the open heart recovery room. The form of this document will vary depending on the corporation. Policies fall within common categories, such as patients, staffing, finance, vendors, and public relations. Within these categories are policies that address common aspects of the business, such as staff hiring, staff performance review, staff discipline, staff promotions, and staff termination. There are sections of policies that are not legislatively clear. It is up to the institution's board of directors to approve these policies. This approval is important, because accrediting and licensing agencies may not judge the merit of the policy as much as the question "did you follow it?"

Some policies are standard throughout the industry, and are sometimes referred to as *boilerplate*, because the corporation uses a model policy developed by an association or outside party and then adopts the model policy as its own. Other policies are designed specifically for the corporation, and are first introduced as a draft written by the president or a vice president of the facility. The draft is presented to the trustees and officers of the healthcare facility for discussion and evaluation. The trustees may ask a committee to analyze the draft and report their suggestions back to the board.

Depending on the bylaws, trustees may discuss the draft policy at a meeting where the general public is invited to attend. The public may or may not be given an opportunity to give trustees input on the proposed policy as per the bylaws. In some cases, a bylaw may give the chairperson of the board the discretion to hear public comment.

If the policy has merit, the draft of the policy is usually given to the legal department, which rephrases the policy to conform to legal standards and clarifies any vague or ambiguous wording. The final version is presented to the trustees for adoption.

Once adopted, the proposed policy becomes official policy of the corporation and is entered into the corporation's policy manual and distributed to the appropriate managers. Think of the policy manual as the corporation's law book. Some corporations make their policy manual available to all officers, managers, and staff, while others may have one policy manual kept in the legal department. The bylaws usually define where the policy manual is kept.

Policies are interpreted by corporate officers and managers who create work rules, or procedures, to implement the policy.

Work Rules and Policy

A *work rule* is a procedure that is an interpretation of a policy and provides guidance to an employee on how to perform their job. For example, the healthcare facility might have a policy of changing an I.V. tube every 72 hours. A work rule may require the staff to label the I.V. tube with the date and time when the I.V. tube was hung, and the date and time when then I.V. tube is schedule to be removed.

Work rules can be defined by any level of management, depending on the management style of the facility. Work rules that govern changing I.V. tubes might be established by the director of nursing, but be based on evidence-based practice research. In contrast, a facility might have a policy that no more than one RN can leave the floor during a shift. The Chief Nursing Officer and the senior management team create work rules for staff leaving the floor, such as for meal breaks. In some hospitals these rules are written into the union contract.

A Nurse Manager can create work rules that are within the scope of their authority. The scope of authority is defined by the Nurse Manager's job description and by their supervisor. Many times, authority is defined in broad, general terms, such as "responsible for the nursing staff of the facility." The Nurse Manager should assume authority for—or delegating to a subordinate—the setting of work rules for tasks that appear to fall within the Nurse Manager's authority.

Works rules may or may not require formal approval from the Nurse Manager's supervisor, depending on the working relationship between the Nurse Manager and their supervisor. In many instances, before instituting new work rules, the Nurse Manager should obtain formal or informal approval for work rule changes that might have an upsetting effect on the staff or on the operation of the facility. No approval is typically required for minor work rules. For example, a work rule that requires LPNs to provide patient nighttime care due to a shortage of nursing assistants is likely to have an upsetting effect on LPNs, especially if only LPNs on the Nurse Manager's unit are affected by the work rule. Even if creating this work rule is within the scope of the Nurse Manager's authority, the Nurse Manager should at least seek the supervisor's informal approval.

On the other hand, a work rule that stipulates that at the beginning of the shift each staff member is given a time slot to take their break, is a work rule that can be initiated by the Nurse Manager without formal or informal approval from the supervisor.

When a Policy Is Not a Policy

A policy may not be valid, even if the board of trustees approves it, if the policy is superseded by other rules. For example, a policy that permits nursing assistants to administer medication is not valid if state law specifies that only LPNs and RNs can administer medication. Also, medication administration is not a task that can be delegated by RNs to unlicensed personnel.

Although conflicts such as this are resolved when the attorney for the facility reviews the final draft of the policy, there are situations in which the law changes after the policy was adopted. That is, the policy was in compliance at the time that it was adopted, but is no longer compliant.

Policies are particularly vulnerable to being superseded, because laws can be modified by any state or federal legislative body. It is common that associations and outside services keep a facility's senior administrators and attorneys abreast of changes in the law so that the facility's policies can be updated.

A policy can be superseded by a collective bargaining agreement with a union and by past practices. A *collective bargaining agreement* is a contract between the facility and a group of employees such as nurses. Terms of the contract might supersede policy. For example, a policy might state that employees must report an hour before their shift begins, but the collective bargaining agreement may say that nurses must report no earlier than a half hour before their shift. The collective bargaining agreement supersedes the policy. And the policy will change. Typically, the contract has verbiage that addresses conflicts between pre-agreement and post-agreement policies.

Enforcing Policy

A policy is the official position of the facility on an issue. Employees and affiliates, such as physicians, agree to uphold policies as a condition to continuing their relationship with the facility. Violating policy is, in essence, a breach of an agreement with the facility, which makes the agreement "voidable" by the facility. *Voidable* means that the facility has the option to terminate the relationship. Alternatively, the facility can impose a penalty that is less than termination. The employee or affiliate then has the option to accept the penalty or terminate the relationship.

The Nurse Manager is in the best position to enforce the facility's policies, because the Nurse Manager administers the day-to-day operations. In doing so, the Nurse Manager has wide discretion on how policies are enforced, much the same as a police officer has wide discretion when enforcing the law. This means that a Nurse Manager can point out an employee's violation of a policy and choose whether or not to take further action, such as giving a warning or taking formal action against the employee.

Any action taken by the Nurse Manager must consider the policy that the employee violated, the employee's character, and the impact the violation has on patients, fellow employees, and the facility. A more formal action is necessary if the violation places a patient, a fellow employee, or the facility in jeopardy, such as failing to administer medication on schedule. Less formal action is appropriate for minor offenses, such as if an employee is a half-hour late for their shift for the first time.

BE CONSISTENT

Policies should be enforced consistently. Otherwise, mixed messages are sent to employees, implying that one employee is favored over another. However, being consistent is no guarantee that a riff among employees won't develop, because consistency is more perception than fact.

Enforcement of a policy occurs privately between the Nurse Manager and the employee because it is a personnel matter. The Nurse Manager gathers facts and then discusses the situation with the employee before taking a course of action. Other employees are likely privy to some facts and not others, which can lead to a misperception of what occurred.

For example, a new nurse has been working for the facility for two months and then takes a two-week vacation. According to policy, an employee must work a full year before taking a vacation. The Nurse Manager does not take action. Nurses in the unit discuss among themselves who the nurse knows in order to receive special treatment. The reality is that the nurse was given a two-week unpaid leave of absence as part of a pre-employment agreement with the facility.

DOCUMENT YOUR ACTIONS

It is critical that a Nurse Manager maintain good records if they plan to take formal or informal action against an employee for violating a policy. Informal violations should be documented in a file memo. A *file memo* is a note to yourself that describes the date and time of the incident, facts that led you to believe there was a policy violation, and evidence that proves the employee committed the violation. A file memo should also include actions that the Nurse Manager took to remedy the situation and the employee's response to those actions.

A file memo documents the event and can be referenced if the Nurse Manager's actions are challenged. Let's say that a nurse is late for the sixth time and the Nurse Manager decides to formally enforce the attendance policy. The result is a hearing with a human resources executive. The nurse says this was the first time she was late. File memos of previous incidents prove otherwise. It would be the Nurse Manager's word against the tardy nurse's if the file memos did not exist.

Be aware that, as a Nurse Manager, your actions—even inconsequential informal enforcement of a policy—could be challenged by the employee in a court of law. Therefore, your actions should be documented so that those documents can be used as evidence in a legal action. Lack of enforcement of a policy can itself be a policy violation and expose the Nurse Manager and the facility to legal actions brought by state and federal government agencies. For example, failing to ensure that a nurse files an incident report after giving a patient wrong medication can result in an action against the Nurse Manager. The Nurse Manager's inaction in the incident is against the facility's policy, and the facility may not support the Nurse Manager's defense.

HANDLING APPEALS

Work rule and policy violations that result in disciplinary action may be appealed by the employee if the health care facility has a formal appeals policy. An appeals policy specifies the procedures an employee must follow to have their case reviewed by a person who has the authority to reverse the previous action. Many facilities have an employee grievance committee. In the absence of such policies/committees, an employee should seek resolution from the human resources department.

The procedure for appealing an action is contained in the grievance section of a collective bargaining agreement. The appeal is referred to as a *grievance*, and is heard in a grievance hearing by one or more hearing officers, who may be senior management of the facility or a third party, such as an arbitrator. A grievance hearing takes a form similar to, but less formal than, a hearing before a judge. If the organization is not unionized, the employees should go to someone in the human resources department. There is a plaintiff and defendant. Each can have a representative, who may or may not be an attorney. The plaintiff is the facility bringing the charge against an employee and representing the Nurse Manager, who determined that the employee committed the violation, and who invoked disciplinary action against the employee. The defendant is the employee who will defend himself against the charge.

In accordance with terms of the grievance section of the collective bargaining agreement, the employee submits written notice of the grievance to the appropriate parties within the facility. The grievance specifies facts in dispute. For example, the employee might feel they were wrongly accused, or that the penalty was inappropriate.

At the grievance hearing, the Nurse Manager presents the reasons for accusing the employee and explains why the imposed disciplinary action was justified. The employee presents reasons why the actions by the Nurse Manager were inappropriate. For example, the employee may claim that the Nurse Manager misinterpreted the policy when creating the work rule that the employee is accused of violating. Alternatively, the employee may claim that, although he or she committed the violation, the disciplinary action imposed by the Nurse Manager was more severe than the discipline imposed on other employees who committed the same or similar violations.

Both sides can present evidence, such as documents and witnesses. Both sides might have the right to question witnesses. After the grievance is heard, the hearing officer reviews the policy, weighs the evidence, and then announces a decision. In some situations, the hearing officer's decision is final. Other times, the employee may have the right to request a hearing before the American Arbitration Association, a nonprofit provider of alternative dispute resolution. The employee might also be able to bring legal action in the courts.

Whether or not an employee can take the appeal beyond the facility depends on the nature of the violation, the nature of the disciplinary action, and state and federal law. The HR department of your facility can advise you of employees' right to appeal.

KNOW WHAT POLICIES EXIST

It is critical that the Nurse Manager be knowledgeable about policies that govern their actions and those of the staff. Review the Nurse Practice Act of your state. The Nurse Practice Act takes precedence over policies adopted by the board of trustees of your facility. The Nurse Practice Act is a statute that basically specifies what a nurse can and cannot do. It sets standards for licensure, regulates nursing schools, establishes continuing education requirements for relicensure, and specifies who investigates violations of the nursing practice. Ignorance of the policies is not an excuse that is defensible in a court of law or before hospital management. Knowing the policies is part of being a good Nurse Manager. Making sure that staff has access to the policies is part of being a better Nurse Manager.

A Nurse Manager should obtain a copy of the facility's policy book and review policies that affect how to manage his or her area of the facility. Policies may use vague or ambiguous terms, such as "the staff should act professionally at all times when on the facility's grounds" but not define what "act professionally" means. Acting professionally is therefore subjective. A Nurse Manager should ask the HR department to clarify any vague or ambiguous terms in writing.

A Nurse Manager should also ask the HR department about previous violations of policy and how those violations were handled. This is especially important to know if the staff is covered under a collective bargaining agreement, because actions in previous cases set precedence for how future violations are handled.

Requests for advice should be made in writing to the HR department. Typically, the HR department responds in writing, providing documentation on which to base future actions. If the HR department prefers to provide advice in-person rather than in writing, the Nurse Manager should follow up with an e-mail that details the advice and asks for confirmation that their understanding is correct. The HR department is likely to respond in an e-mail confirming that the Nurse Manager understands the advice given, or clarifying any misunderstandings.

POLICIES AND POLITICS

Although policies are supposed to be created to address situations that could have an adverse effect on patients, employees, or the facility, there are times when policies are created solely for political purposes.

To some, the term "politics" brings to mind images of backroom, unethical deals resulting from payoffs. Although scandals such as these have surfaced in the press, "politics" does not have a good or bad connotation. *Politics* are the social relations among a group involving authority or power evolving into a process for making decisions for the group.

The political process begins with a problem that a group of people faces, such as providing healthcare for the community. A member of the group has an idea to solve the problem, but needs the group to give authority to enact the solution. Politics is used to gain this authority.

A group grants authority formally, through election or appointment, and informally, by looking to a person for advice and leadership. Authority is sometimes granted to a person by members of a group based more upon perception than fact. Each member establishes criteria for granting authority. For example, some may expect the person to be slim, fit, and neatly dressed in a suit. Others may want the person to be "one of them," coming from a similar background and having the same economic status as they do. The person seeking authority creates the perception that he or she meets the majority of the group's expectations.

Politics doesn't stop once a person gains authority, because a policy to solve the group's problem must be accepted by the group. Therefore, the policy must contain elements that members feel will address what they perceive to be the problem.

A proposed policy can serve two purposes. First, the policy must address the real need of the facility to deal with patient, employee, or facility problems. Second, the policy can address perceived problems identified by members of the board of trustees and their constituency. A perceived problem may not exist, but takes on the importance of a real problem—and at times a perceived problem is more important than a real problem.

Suppose the hospital's free clinic is perceived by some in the community as giving second-rate healthcare because patients are rarely seen by physicians who see paying patients. Clinic patients are usually examined by a nurse practitioner and then seen by a resident physician only if the patient's condition warrants it. The nurse practitioner and the resident physician provide the patient with excellent healthcare. However, for political purposes, the chief operating officer might propose a policy that requires every physician who has admitting rights to donate one week of time over the course of a year to treating patients in the clinic. In this case, the policy addresses a perceived problem rather than a real problem.

Politics also plays a crucial role in day-to-day operations, when the Nurse Manager and the staff interact with other departments within the facility. No Nurse Manager wants their authority subverted by another employee, particularly another Nurse Manager. It is therefore important for a Nurse Manager to make sure their actions are not seen as formally or informally intruding upon another Nurse Manager's authority.

Suppose a Nurse Manager devises a solution to a problem that affects other areas of the facility as well as the Nurse Manager's own area. A prudent approach is to informally discuss the solution with other managers before presenting a proposed policy to senior management. This reduces the likelihood that other managers will perceive the Nurse Manager's actions as a move toward acquiring more authority.

Policy and Power

Policies are developed from a position of power. Power is the ability to influence others to achieve a goal. The goal of a policy is to influence the behavior of the staff and others affiliated with the facility to address an issue confronting the facility.

TYPES OF POWER

There are six ways power is used to influence others:

- **Expert power**   Knowledge, such as a physician's advanced training, which gives them the ability to influence the actions of nurses.
- **Reward power**   The use of rewards or favors to influence a person, such as promoting a nurse to charge nurse for exceptional performance.
- **Information power**   Occurs when a person has information that others need to know, such as a nurse showing other nurses how to use a new I.V. pump.
- **Coercive power**   A person has the ability to punish others, such as a Nurse Manager disciplining a nurse for a policy violation.
- **Connection power**   A person associates with others who have authority, such as a Nurse Manager seen having lunch with the chief operating officer of the facility.
- **Legitimate power**   Gained by a person's position in the facility, such as the nurse's appointment to the position of Nurse Manager.

GAINING POWER

To effectively develop policies and procedures, the Nurse Manager must be powerful so that their ideas are acceptable to senior management and staff. There are two factors that influence whether or not a Nurse Manager gains power: attitude and behavior.

The Nurse Manager must present an outward appearance of self-confidence by establishing the impression of being in charge. This begins by developing a powerful self-image and giving a general appearance that you are a problem solver. Body language is critical. Maintain eye contact and be confident in your gestures and movement. Speak in a confident voice, using good grammar and diction. Be sure to use appropriate vocabulary to ensure that your thoughts are understood.

Make others feel good about meeting with you. Be honest. Acknowledge your mistakes and be a graceful winner and loser. Credit others for their contributions. Promise only what you can deliver, and follow up on those promises. Always resolve conflicts in the best interest of the patient, but be sure that all parties to the conflict save face.

Create a network of contacts inside and outside the facility who are good sources of information and who can provide advice and support. The network helps to expand your area of influence. Seek a mentor who can steer you in the right direction. A mentor is usually a person who has gained power and is willing to guide up-and-coming managers.

Set goals for your area and make those goals known to your staff, to other managers, and to senior managers. Develop a strategy to achieve those goals, and then share the results of your achievements with others. Even partial success illustrates your technique for being a pragmatic problem solver.

Develop expertise in areas that are important to senior management. Take courses, if necessary at your own expense and on your own time, that lead to certification in specialties related to your unit or that are lacking in your facility. You will be looked upon as the expert within the facility. This book is a great start in gaining the skills necessary to becoming a good Nurse Manager.

Keep high visibility within the facility. Take on leadership roles in committees and activities that are important to senior management. This demonstrates your commitment to the facility and the breadth of your skills to senior management.

As a Nurse Manager, empower your staff to make decisions without having you approve them. Set forth your expectations, and then permit your staff members to devise and enact their own solutions to problems as long as your expectations are met.

Summary

A policy is a rule of practice that supplements the bylaws of a for-profit or not-for-profit corporation. Policies are the day-to-day rules that govern how managers, affiliated healthcare providers such as physicians, and staff care for patients and operate the healthcare facility.

A bylaw is a basic rule for the conduct of the corporation's business and affairs, established at the time that the company is formed. A policy is created according to procedures described in the bylaws of the corporation, and typically requires adoption by the board of trustees. Facilities can adopt a model policy developed by an association or outside party that covers policies that are standardized throughout the healthcare industry.

Once a policy is entered into the facility's policy manual, it is distributed to corporate officers and managers who are responsible for the implementation of the policy. Corporate officers and managers interpret policies into work rules, or procedures, to implement the policy. A work rule is a procedure that is an interpretation of a policy and provides guidance to an employee on how to perform his or her job in compliance with the policy.

A policy may not be valid, even if the board of trustees approves it, if the policy is superseded by other rules. Policies are particularly vulnerable to being superseded by legislation. A policy can also be superseded by a collective bargaining agreement and by past practices.

Violating policy is a breach of an agreement the employee has with the facility, which makes the agreement voidable by the facility. Voidable means that the facility has the option to impose a penalty on the employee.

The Nurse Manager has some discretion on how policies are enforced. Policies should be enforced consistently so that mixed messages are not sent to employees. Work rules and policy violations that result in disciplinary actions may be appealed by the employee.

There are times when policies are created solely for political purposes. Politics are the social relations among a group involving authority or power evolving into a process for making decisions for the group. Policies are developed from a position of power. Power is the ability to influence others to achieve a goal.

Quiz

1. What is a grievance?
   - (a) A complaint by a Nurse Manager against an employee
   - (b) A complaint by a Nurse Manager against a member of a collective bargaining unit
   - (c) An appeal of disciplinary action again a member of a collective bargaining unit
   - (d) None of the above

2. What is a policy?
   - (a) A supplement to a bylaw
   - (b) A supplement to a work rule
   - (c) A type of work rule
   - (d) A type of bylaw

3. In which way can others be influenced?

    (a) By a person's knowledge

    (b) By use of rewards or favors

    (c) By having information that others need to know

    (d) All of the above

4. A policy supersedes the Nurse Practice Act.

    (a) True

    (b) False

5. Policy is the ability to influence others to achieve a goal.

    (a) True

    (b) False

6. The board of trustees is granted the right to adopt policies by:

    (a) Work rules

    (b) Bylaws

    (c) Another policy

    (d) All of the above

7. An employee may challenge enforcement of a policy in court.

    (a) True

    (b) False

8. Policies can be proposed by:

    (a) The board of trustees

    (b) Senior management

    (c) Nurse Managers

    (d) All of the above

9. A policy must be distributed to all managers for the policy to take effect.

    (a) True

    (b) False

10. A policy must be written.

    (a) True

    (b) False

CHAPTER 7

Legal Issues

In a society that has become more and more litigious, the Nurse Manager must become increasingly aware of laws, rules, regulations, and policies that govern patient care. Failure to do so can result in legal action against the Nurse Manager, loss of license, fines, and in rare incidents, incarceration.

Nurse Managers are not expected to be lawyers. However, they must be fluent in the legalities of healthcare to ensure that precautions are taken to minimize the effect of potential legal action, and to properly handle claims that might be filed against themselves or the healthcare facility.

In this chapter, you'll explore the legal system and laws that govern healthcare. You'll also learn how to prepare for litigation, and steps you need to take to reduce the likelihood that you will become a target of legal action.

Legal Environment

The U.S. legal system has evolved from English common law, introduced to the United States by the settlers. Common law consists of a standard set of acceptable practices that are used to protect individuals and resolve disputes. Common laws are not written or approved by a government body. Instead, they are established by tradition of a community as determined by judges in the court system.

Common law in the U.S. empowered a judge to make rules that gave nearly instant protection to the community from persons whose behavior was seen as detrimental to society. A judge resolved disputes by using common sense based on what he thought was acceptable behavior.

As society grew and communities began to interact with one another, there was a need to formally codify rules and regulations that could be uniformly applied to groups of communities, counties, and states. The first codified rule was a constitution that established the fundamental rules and principles for governing an organization. Communities banded together to form a state and adopted a state constitution that defined fundamental rules for that state, established power and duties of government, and guaranteed rights to residents of the state. States banded together to form the United States, and adopted the U.S. Constitution. No law supersedes the U.S. Constitution.

The U.S. Constitution created three branches of government: executive, legislative, and judicial. The legislative branch creates proposals that become law when approved by the majority of the legislators, and by the executive branch in the form of a presidential signature. The executive branch enforces the law, and the judicial branch administers the law.

CIVIL LAW AND CRIMINAL LAW

There are two categories of law: civil and criminal. Civil law governs how disputes between individuals and entities, such as corporations, are resolved. Claims of violating a civil law can be brought to the lowest civil court in a state by any person or entity that believes an injustice was perpetrated. The person bringing the charge is called the *plaintiff*, and the person who is alleged to have violated the civil law is called the *defendant*. The plaintiff must present to the court evidence of their claim against the defendant. The defendant has the right to challenge the evidence. The judge or a petit jury make a decision based on the preponderance of the evidence. A *petit jury* is an ordinary jury of 12 or fewer persons who determine the facts in a case after hearing both sides. Only monetary penalties are imposed for violating a civil law.

Criminal laws governs what is unacceptable behavior in society. The plaintiff in a criminal matter, is called the prosecutor, and is sometimes referred to as the state. Criminal law falls into one of three categories of crime. These are disorderly person, misdemeanor, and felony, which is sometimes referred to as a high misdemeanor. Disorderly person is the lowest degree of crime, and felony is the highest.

Formal accusation of violating a misdemeanor or felony law is given by community members who come together and form a grand jury. A grand jury listens to purported facts presented by the prosecutor, and then decides if a crime has been committed, and if the defendant likely has committed the crime. The grand jury does not determine if the defendant did commit the crime, because that's the job of the petit jury. If the grand jury determines that a crime has likely been committed, then the grand jury issues an indictment. An *indictment* is a formal statement that says, "we the grand jury, accuse you, the defendant, of committing a specific misdemeanor or felony." The defendant is then put on trial. If found guilty, the defendant could receive a monetary fine and/or be incarcerated as a penalty for violating the criminal law.

CASE LAW

Laws are often written broadly in terms that convey the wishes of the legislators. Judges then interpret the law. A judge's interpretation sets a precedent that applies to other similar cases within the judge's jurisdiction. This is referred to as case law.

A judge may misinterpret a law. Therefore, the plaintiff or the defendant can appeal the judge's ruling through an appeals process that can lead all the way to the United States Supreme Court. A judge's ruling becomes case law for the judge's jurisdiction unless the ruling is overruled by an appeals court, at which time the appeals court's ruling takes precedent as case law for the jurisdiction of the appeals court.

Lawmakers can revise a law if they disagree with the court's interpretation of it, and thereby can overrule case law.

THE LEGAL PROCESS

Nurse Managers, nurses, and other healthcare providers have legal exposure that is not common to the average person, because healthcare by its nature is invasive and may cause some degree of harm to the patient in order for the patient to resume activities of normal daily life.

For example, rehydrating a patient with an I.V. can, in extreme cases, be considered an atrocious assault and battery. An assault is a threatened or attempted physical attack by a person who is able to cause bodily harm. A battery is causing bodily harm. "Atrocious" sometimes refers to the degree of harm, such as causing a person to bleed. And, should an inappropriate I.V. fluid be used that causes the patient to die, the violation could escalate to atrocious assault and battery with a deadly weapon and manslaughter. Charles Cullen, referred to as "The Angel of Death," was a healthcare worker who was administering a potassium I.V. to induce cardiac arrest and death in patients. Potassium is necessary to the body, but Cullen possessed knowledge of how to use it to cause harm. An unsuspecting healthcare worker could accidentally infuse a potassium I.V. too quickly in a patient. The lack of knowledge concerning the consequences does not lessen the outcome for the patient, or the legal action for the healthcare worker.

Aside from criminal law violations, healthcare professionals are also exposed to civil action if they do not provide normal and customary care to the patient and, as a result, the patient experiences a loss. The courts award a patient monetary compensation for losses caused by a healthcare provider giving less than normal and customary care. The definition of the term "normal and customary care" is a turning point in many civil cases against healthcare providers. Experts are asked to testify whether or not the healthcare provider's actions were normal and customary.

PATIENT'S LOSS

The patient's loss is another contentious factor in these cases. Less than normal and customary care does not, on its own, merit monetary compensation. The action must have caused the patient to lose something, such as the ability to continue activities of daily life (pain and suffering), incurring additional medical expenses, or losing wages.

Let's say that hospital policy requires the I.V. tubing be changed every 24 hours, but 48 hours goes by before the tubing is changed for a patient. The patient is not adversely affected. In this example, the patient received less than normal and customary care, but did not experience a loss because he or she did not have an adverse reaction. The patient potentially could have suffered loss if infection resulted from the tubing not being changed on time, but the loss never materialized.

Medical treatment is risky, and it is not unusual for a patient to suffer a loss as a result of treatment. Even if a patient accepts a particular risk before beginning treatment, the patient may still seek action against the healthcare provider if the patient experiences a loss. For example, a patient is given medication for high blood pressure and is told that impotence is a potential adverse side effect of the medication. The patient accepts the risk and takes the medication. Six months later he is impotent, contacts his attorney, and files action against the healthcare provider, healthcare facility, and the pharmaceutical manufacturer. Although the patient may not prevail in court, he still can file a lawsuit.

THE RESPONSE

Civil complaint can be taken even if the claim is without merit. Once the complaint is filed, the defendant has the option to respond or ignore the action; however, not responding typically signals to the court that the defendant does not challenge the claim. This results in a default judgment against the defendant.

Therefore, a response must be filed to all actions, even those where the plaintiff's case is obviously flawed. A response is filed by the defendant's attorney or the attorney for the carrier who covers the healthcare provider with malpractice insurance.

SETTLEMENTS

The plaintiff, the defendant, and the defendant's insurance carrier weigh the expense of litigation against the potential award if the case is successful for the plaintiff. It is not unusual for all sides in the case to arrive at an amiable agreement, referred to as a *settlement*.

It is important to understand that many settlements in civil actions are made for economic reasons, and not because the healthcare provider provided less than normal and customary care to the patient. Some healthcare providers find such settlements difficult to accept, because the patient was given normal and customary care and any loss was a risk the patient accepted in writing before receiving care. However, the healthcare provider may have little influence on whether or not a case is settled. Typically, the malpractice insurance carrier agrees to pay claims if the healthcare provider agrees to have the insurance carrier represent them in the case. The insurance carrier may simply determine that settling the case would be cheaper than litigation, and thus settle the claim regardless of fault.

THE LEGAL PROCESS AND YOU

The legal process is a paradox. It is predictable because the process is clearly defined by law, yet it is unpredictable because judges apply the law to fit the unique facts of each case. There are two ways in which a Nurse Manager becomes personally involved in the legal process: as a witness or as a defendant.

A *witness* is someone who can provide facts related to a criminal or civil action against a defendant. Facts provided by a witness can become evidence if they help prove culpability or innocence. A witness is typically asked to discuss all or a portion of the case with the plaintiff's attorney, defendant's attorney, or both, in a semiformal setting. The attorney might visit your place of work or ask that you meet in their office.

If facts that the witness provides help the plaintiff's or defendant's case, the witness is likely to be subpoenaed to formally give facts under oath. A *subpoena* is a legal document that requires that a person appear in a legal forum. Sometimes a subpoena requires a person to deliver documents to an attorney so that the attorney can review them as part of a legal action.

In a criminal proceeding, the subpoena is for the witness to present their facts before a grand jury. In civil action, the witness presents facts in a deposition.

A *deposition* is a legal forum outside of court where attorneys for the plaintiff and the defendant question a witness. The witness is under oath to tell the truth and the proceeding is documented by a court reporter. Either the plaintiff's attorney or the defendant's attorney issues the subpoena for the witness to appear at the deposition. In the deposition, that attorney asks questions of the witness, which is referred to as direct examination of the witness. Then, the other attorney can also ask questions, which is known as cross examination of the witness. Statements made by the witness become evidence in the case and can be used in further legal proceedings.

You must respond to a subpoena. Failing to do so gives the attorney the option to have you arrested and placed in jail until you respond. It is best to consult your attorney on how to respond to a subpoena. If the subpoena is for documents related to a patient, then immediately give the subpoena to the attorney for your facility. Usually the attorney will handle matters directly.

Maintaining patient confidentiality is of utmost importance when you are approached to provide information to an attorney either informally or in a formal legal forum. The attorney must demonstrate a legal right to access patient information before you discuss anything about the patient, or provide any patient records to the attorney. It is always best to consult the healthcare facility's attorney when approached by an outside attorney. The facility's attorney can advise you regarding what you can and cannot mention about the patient and about the situation in question.

YOU ARE SUMMONED

Legal action against you begins when you receive a summons. A summons is different from a subpoena. A *summons* is a legal document telling you to appear in court to answer charges that someone has filed against you. A subpoena is a legal request for you to appear in a legal forum, which might be a court room, to provide facts for a case. You cannot ignore a summons. If you do so in a criminal proceeding, then you face possible arrest. If you do so in a civil proceeding, you could be found liable by default. Immediately consult an attorney if you receive a summons, because you must begin preparing to defend yourself against the charges.

It is common for several defendants to be named in a lawsuit, especially medical malpractice lawsuits. Typically, the plaintiff's attorney names other workers, the facility, equipment manufacturers, and others as defendants. The theory is that each is responsible for a percentage of the plaintiff's loss. In a multidefendant civil lawsuit, the jury is asked to determine the monetary loss to the plaintiff, if any, and then determine the culpability of each defendant, which is assessed as a percentage. The percentage is applied to the monetary loss to determine how much each defendant pays the plaintiff.

Don't assume the following if you are sued:

- Your exposure to legal liability is lessened by the fact you are one of many defendants in the case. It isn't.

- You are not potentially personally liable because you were working for the healthcare facility at the time of the incident. You can be held personally liable.

- The attorney for the healthcare facility will defend you. An attorney is obligated to defend his or her client. The client of the attorney for the healthcare facility is the healthcare facility—not you. There may be conflicts between the interest of the healthcare facility and a Nurse Manager. For example, the Nurse Manager may have violated a policy of the healthcare facility that resulted in a patient's loss, in which case the attorney for the healthcare facility is likely to try to minimize the facility's culpability by saying that the Nurse Manager violated the facility's policy. The Nurse Manager needs his or her own attorney to argue, for example, that the facility failed to hire a sufficient number of nurses, making it nearly impossible to comply with the policy. This is also why it is important to have your own malpractice insurance, even if the institution carries it.

BE PREPARED

It is not uncommon that action is brought months and even years after the patient has left the facility. As a result, you may have no recollection of the patient unless the events giving rise to the case were extraordinary. You'll need to review the case. This is why thorough documentation is important, as chances are you won't remember the details of the case.

Begin by reviewing the patient's medical record. Pay particular attention to notes and other documentation written around the time of the event that led to the action, especially those written by you.

Try to reconstruct the event based on facts that you find. Avoid postulating on what you think happened, because your recalculation might be inaccurate. Do not make any assumptions. It is easy to fall into the trap of basing your reconstruction on what normally would have happened, because maybe the response to the event was different this time.

Create a timeline of the event by piecing together factual information from hospital records and personal notes. Identify the source of facts listed on the timeline. Don't be afraid to leave blanks in the timeline if you are unsure of what happened during that period.

Examine incident reports filed around the time of the event, including incidents from other areas of the facility. For example, your unit might have had a faulty piece of equipment that was borrowed from another unit. It may turn out that staff from the other unit wrote an incident report regarding the equipment, but no one on your staff wrote one.

GIVING TESTIMONY

The time will come when you are questioned in a formal legal proceeding by an attorney. Your attorney should have prepared you for giving testimony by telling you what kind of information can be given and how to give it.

You'll be invited into the legal forum, where you will be sworn in by an officer of the court, such as the court clerk, the judge, or an attorney. After you have a moment to settle into your chair, an attorney begins asking you questions. The attorney will use different demeanors when asking questions in an attempt to elicit a desired response from you. For example, the attorney may be loud and threatening, asking short, pointed questions to shake your testimony. Or the attorney may be extremely friendly, to make you feel comfortable saying just about anything. Your job is to ignore how questions are asked and simply focus on the question and your response.

The attorney must have a legal foundation for asking a question. In establishing that foundation, the answers to some questions that you are asked may seem obvious, causing you to wonder why the attorney asked it. For example, the attorney might ask you if you were in charge of the XYZ unit at the time of the event, even though the attorney already knows you were. The reason for the question is that the attorney needs for you to say so in order to establish the foundation to ask follow-up questions, such as whether the plaintiff was a patient under your care during the incident.

The following are some pointers to keep in mind if you are ever questioned in a formal legal proceeding:

- Listen carefully to each question. Sometimes the attorney wants you to provide specific facts but not draw a conclusion. For example, instead of asking you to describe the patient's condition, the attorney may ask you to provide facts such as vital signs.

- Answer only the question that was asked. For example, the attorney might be exploring the patient's therapeutic level of heparin and ask you for the patient's PT value. You know that the attorney probably should have asked for the patient's APTT value. However, you should only answer the question that was asked of you.

- Ask if the attorney can restate the question if you don't understand it.

- Simply say, "I don't recall" or "I don't know" if you are unable to answer the question.

- Say "I'm not qualified to answer that question," if the answer requires you to go beyond your expertise. For example, the attorney might ask if the patient was depressed. Depression is a medical diagnosis made by a physician, which is beyond your expertise.

- You can review your notes to respond to a question; however, ask permission to do so before looking at the notes. If you are not given permission, then say that you are unable to answer the question.

- Some attorneys may ask you to explain the event in your own words rather than ask pointed questions. You should respond only with facts. Don't give an opinion, draw a conclusion, or embellish the facts. If you told a nurse to change an I.V. and didn't verify that the I.V. was changed, then you can't say that the nurse changed the I.V. You can say that you instructed the nurse to change the I.V., and that you didn't verify that the I.V. was changed because the nurse was validated to change an I.V.

- Refer to the document that was written at the time of the event rather than provide your recollection of the event. You can say to the attorney: "I did what I wrote in the record."

- Remain calm, composed, and truthful throughout your testimony.

Legally Prepared

The success of a legal action is frequently determined by ongoing documentation of patient care by the healthcare facility's staff. Documentation includes the patient's chart, lab results, prescriptions, nurses' notes, treatments, and other information that describes the who, what, when, where, and why of patient care. That is, who gave what care when? Where was care given, and why was care given. The documentation should tell the story of the patient's care, including facts and opinions given by qualified experts.

In addition, operational activities should be documented, such as staff schedules, staff training, compliance with license requirements, facility and equipment maintenance, facility and staff certifications, and other information that attests that the facility is operating safely within legal requirements.

INCIDENT REPORTS

The Nurse Manager must also ensure that a detailed incident report is written and filed immediately when an unusual event occurs such as a medication error, a patient falling, or equipment failure. The incident report tells the facts that led up to the event, facts about the event, and actions taken after the event.

The incident report should include typed statements from those who were involved in the incident, including the person directly responsible for the incident, and from those who were associated with the incident. Let's say that the wrong medication was given to a patient. The medication was a verbal order taken over the phone by two nurses who wrote the order in the patient's chart. The patient's primary care nurse prepared and gave the medication to the patient. All three nurses must give statements for the incident report.

Statements should be written as close to the time of the event as possible in order to capture the best recollection of the event. The Nurse Manager should ask the staff to write their statements by hand and to sign the statements. Later, an assistant can type the official statement and give it to the risk management department staff to review, make corrections, and eventually sign. Statements should not be a collaborative effort, but rather should reflect each staff member's independent observation of the facts related to the event.

Incident reports are submitted to the appropriate department of the healthcare facility, which then analyzes the facts and possibly conducts an independent investigation to determine how similar events can be avoided in the future. Incident reports do not belong in the patient's chart.

Nurse Practice Acts

Many legal issues arise from rules created by the State Board of Nursing and, in particular, the Nurse Practice Act for each state. The State Board of Nursing and the Nurse Practice Act define requirements for practicing nursing in a state.

The Nurse Practice Act sets educational requirements of licensure, authorized titles for nurses, and defines the health care nurses can provide to patients. The state Board of Nursing implements and enforces the Nurse Practice Act.

It is the responsibility of a Nurse Manager to be sure that each nurse working on the Nurse Manager's unit has a license to practice in the state. Nurses licensed in other states can practice nursing if they receive an endorsement from the State Board of Nursing, which confirms they have a valid nursing license from another state.

Conflicts among Nurse Practice Acts can lead to unexpected violations. For example, a conflict can involve case managers who manage patients in a healthcare facility located in one state, but the patients live in a neighboring state. Technically, the case manager must be licensed in both states.

Exposure to Legal Claims

Patients have the right to receive appropriate treatment that they approve, and the right to confidentiality throughout their treatment. When these rights are violated, the patient can file a civil, and possibly a criminal, complaint against their health providers.

There are several critical areas where nurses are vulnerable to this type of legal action. These include malpractice, negligence, delegation, informed consent, and confidentiality.

MALPRACTICE

Malpractice occurs when a nurse is negligent while caring for a patient, and as a result the patient is injured. Nurses, as well as other healthcare providers, are expected to care for a patient in the same manner as any reasonable and prudent nurse would care for a similar patient. The injury can be caused unintentionally, such as giving a patient the wrong medication as a result of misreading the drug label, or intentionally, such as inserting a saline lock that later becomes infected. The insertion was intentional, but the infection was not. Malpractice claims are brought in civil court.

NEGLIGENCE

Negligence is the failure to act in a way that any reasonable and prudent person would act in a similar circumstance. Negligence applies to all persons, including professionals, whereas malpractice applies only to professionals. For a nurse, it might be negligent to carry hot coffee in an uncapped cup and bump into a patient in the hallway, spilling the coffee on the patient and causing the patient to receive first-degree burns. A reasonable and prudent person would anticipate the possibility of spilling the coffee, and thus cover the container while walking through the halls. Negligence claims are also brought in civil court.

DELEGATION

Delegation occurs when a nurse assigns another healthcare worker to perform patient care. The nurse is accountable for the actions of that healthcare worker. Both the healthcare worker and the nurse can face malpractice or negligence claims should injury occur to the patient. The nurse is obligated to know whether the healthcare worker has the skills necessary to safely perform the delegated patient care. Furthermore, the nurse must verify that care was properly given to the patient.

INFORMED CONSENT

Informed consent must be given by the patient before the patient receives treatment. Before informed consent can be given, the patient must understand the physician's explanation of the risk and benefits of the treatment. The nurse is obligated to alert the physician if the patient questions the treatment either before or after signing the informed consent. The nurse is exposed to a malpractice claim if he or she has reason to believe that the patient does not fully understand the risk and benefits of the treatment after signing the informed consent, and the nurse does not alert the physician.

It is the surgeon's or physician's responsibility to obtain the proper consent. The nurse does not explain the treatment, but witnesses that the signature is indeed the patient's. Laws regarding informed consent differ among states. Some states require informed consent for every treatment, such as insertion of a saline lock. Other states require an informed consent for treatment that could result in serious harm to the patient. In addition, there are different rules for treating patients who are unconscious in emergency and nonemergency situations. Always consult with the attorney for your healthcare facility for guidelines for implementing informed consent.

CONFIDENTIALITY

Patient information must be kept private and shared only with healthcare professionals who are directly involved in the patient's care. Confidential information can be shared if doing so protects the safety of others. Failure to keep patient information confidential exposes the nurse to a malpractice claim.

Living Wills and Durable Power of Attorney

A living will is a written document created by the patient that specifies the care the patient wants to receive if he or she is unable to communicate, for whatever reason, to the healthcare provider. For example, the living will might say, "Do not keep me alive by artificial means," "Do not resuscitate me," "Do not place me on a feeding tube," "Do not start dialysis," or "Do not connect me to a respirator."

It is difficult for people to list all the possible ramifications of illness and to include guidance related to them all. Thus, a durable power of attorney for healthcare (DPAHC) designates one person to make all healthcare decisions for the patient in case the patient is incapacitated. This one person will decide each question as it arises.

A living will or a DPAHC must be executed while the patient is competent. A living will may be relatively informal, just something written and signed by the patient. A DPAHC must be formally created, usually by an attorney, and signed by the patient. The patient's signature is notarized while the patient has the mental capacity to do so.

States use different terminology for these documents, such as the Advance Directive for Healthcare Act, or the Death with Dignity Act. It is therefore very important for you ask the attorney for your healthcare facility for guidance whenever you encounter a living will or DPAHC.

The living will or DPAHC should be included with the patient's chart. Carefully document all conversations with the patient, family, and staff, if the patient's living will indicates that the patient does not want extraordinary life support measures taken. A patient can rescind a living will at any time.

Before withholding life support per the patient's living will the nurse should contact the healthcare facility's risk management department and attorney to be sure that the staff's actions conform to the patient's wishes and the law.

The healthcare provider is bound by the terms of a valid living will or DPAHC, even if the patient's family objects to those terms. For example, a 100-year-old woman who is terminally ill with cancer may state in her living will that she wants extraordinary life support measures taken. Family members—and healthcare providers—may disagree, feeling they don't what to extend her pain and suffering. However, the patient's wishes must be followed.

Health Insurance Portability and Accountability Act

The Health Insurance Portability and Accountability Act (HIPAA) is federal legislation that regulates patients' records, and provides privacy by limiting ways healthcare providers can use patient information.

Healthcare facilities are required to establish privacy procedures that specify who can access patient information, and how that information is accessed, regardless of whether the information is on paper, computerized, or transmitted orally over the telephone.

Violations of the patient's privacy rights expose the person who reveals the information and the healthcare facility to both civil and criminal actions. Expensive fines may be incurred by the facility for any employees failing to comply with HIPAA.

Medical Orders

Medical orders must be written by a physician, nurse practitioner, or physician assistant (PA), and carried out by the nurse. Some states limit the type of medical orders a nurse practitioner or PA can write. The physician, nurse practitioner, or PA is legally responsible for the correctness of the medical order. The nurse is responsible for carrying out correct orders, and for not carrying out incorrect orders.

If the medical order may harm the patient, the nurse should not carry it out, and should immediately call the person who wrote the medical order, informing them of the problem and asking for clarification. For example, a medical order for medication might be contraindicated for the patient. The medication is held, and the physician, nurse practitioner, or PA is contacted.

In some situations, the nurse may accept either telephone medical orders or verbal medical orders. The nurse must document the medical order in the patient's chart, and the physician, nurse practitioner, or PA must sign the order as soon as possible.

Two nurses should listen to telephone medical orders to verify the medical order. The nurse must repeat the patient's name and the medical order before disconnecting the call.

A nurse should refuse to accept verbal medical orders unless there is an emergency, such as a cardiac arrest. In emergencies, document the medical order as soon as possible in the patient's chart and have the medical order signed by the physician, nurse practitioner, or PA.

Personnel Law

The Nurse Manager must be familiar with laws that regulate personnel matters within the healthcare facility, because violations can result in civil action taken by an employee. The HR department of your healthcare facility can provide you with guidelines for adhering to local, state, and federal laws that cover personnel matters.

In particular, you should be familiar with terms of the National Labor Relations Act, which defines unfair labor practices, some of which apply to healthcare facilities. For example, this law prohibits a candidate from being asked their age, religion, whether or not a woman candidate plans to have children, and other personal information that may unfairly disqualify the candidate for the position.

In addition, review with the HR department provisions of the Americans with Disabilities Act that relate to healthcare facilities. The HR staff can help you inadvertently avoid violating this law.

Summary

The legal system has evolved from English common law introduced to the United States by the settlers. Common law consists of a standard set of acceptable practices that are used to protect individuals and resolve disputes. The first codified rule is a constitution that establishes the fundamental rules and principles for governing an organization.

There are two categories of law: civil and criminal. Civil law governs how disputes between individuals are resolved. Criminal law governs acceptable behavior in society. Laws are also made when a judge rules in a legal case. This is referred to as case law.

Legal action begins when a person receives a summons. A summons is a legal document telling you to appear in court to answer charges that someone has filed against you. A subpoena is a legal request for a person or records to appear in a legal forum, which might be a courtroom, to provide facts for a case.

A nurse can be exposed to legal action in several areas. These are malpractice, negligence, delegation of a task to another person, informed consent, and violating confidentiality.

Healthcare providers must abide by a patient's living will, which specifies the care the patient wants to receive if they are unable to communicate for whatever reason, with the healthcare providers.

Quiz

1. A person is a subject of a civil action when they receives a:
   (a) Subpoena
   (b) Summons
   (c) Judgment
   (d) None of the above

2. What kind of law governs acceptable behavior in society?
   (a) Criminal law
   (b) Civil law
   (c) Petit law
   (d) None of the above

3. What action can be taken against a nurse who fails to properly care for a patient?
   (a) Negligence
   (b) Malpractice
   (c) Liability
   (d) None of the above

4. The Nurse Practice Act is the same in every state.

   (a) True

   (b) False

5. A Nurse Manager is not exposed to legal action once patient care is delegated to a licensed nurse.

   (a) True

   (b) False

6. A nurse can be held legally liable if:

   (a) A patient fails to sign an informed consent and treatment is withheld.

   (b) A nurse witnesses the signing of an informed consent and the patient dies as a result of the treatment.

   (c) The nurse proceeds with treatment after learning that the patient doesn't fully understand the risks associated with the treatment after the patient signs the informed consent.

   (d) None of the above

7. The patient's family can override a living will once the patient loses consciousness.

   (a) True

   (b) False

8. A nurse is exposed to civil action if the nurse:

   (a) Carries out an incorrect medical order

   (b) Accepts a verbal medical order in an emergency

   (c) Accepts a medical order over the telephone

   (d) None of the above

9. A patient can verbally change a living will while in the hospital.

   (a) True

   (b) False

10. Statements in an incident report should be a collaborative description of the event.

    (a) True

    (b) False

CHAPTER 8

Healthcare Economics

Do you know why some people are rich—and perhaps you're not? Economics. Do you know why healthcare costs are through the roof? Economics.

You've heard the term economics before, but perhaps not in the context of healthcare because most of us think of them separately. Economics is paying your bills. Healthcare is taking care of yourself, your family, and your patients. In simple terms, *healthcare economics* is the system used to provide and pay for healthcare.

The healthcare industry is undergoing a disruptive economic change as it moves from an employer-based health insurance system to one that gives patients the direct power to influence healthcare prices.

In this chapter, you'll learn how the employer-based health insurance system has disrupted fundamental economic principles that kept prices under control. You'll also learn principles that you can use to manage your area of a healthcare facility through this radical change in healthcare economics.

Basic Economics

In a self-sufficient world, each of us has the capability to fulfill all our needs for food, clothing, shelter, education, entertainment, and other things we require to survive and enjoy life. Realistically, no one is self-sufficient. Things we cannot do for ourselves create a demand for others to do it for us, which is the basis for an economy.

An *economy* is a system of production, distribution, and consumption. In a primitive barter economy, two consumers exchange products or services to fulfill each other's demand. For example, a dairy farmer and hay farmer might exchange a gallon of milk for a bale of hay. The dairy farmer has a demand for hay to feed his cows, and the hay farmer has a demand for milk to feed her family. The value of a product or service is determined by what both parties feel is a fair exchange. In this case the farmer sets the value of a gallon of milk equal to a bale of hay.

The value is influenced by demand for and availability of a product or service. High demand and low availability increase the value of a product or service. Conversely, low demand and high availability decrease the value. Suppose there are one dairy farmer and two hay farmers. The dairy farmer exchanges a gallon of milk for a bale of hay with the first hay farmer. The dairy farmer no longer has a demand for hay when the second hay farmer arrives for a gallon of milk. The value of hay has dropped to zero, because the dairy farmer isn't going to exchange a gallon of milk for hay that he doesn't need.

An economy is fluid and can change unpredictably based on economic influences. For example, the second hay farmer still has a demand for a gallon of milk, so he tries to find another need to fulfill for the dairy farmer in exchange for a gallon of milk. He looks around the dairy farmer's property and notices a fence that needs mending. He proposes to fix the fence in exchange for a gallon of milk. If they agree, then the economy expands. Originally the economy supported two products—milk and hay. Through the initiative of the second hay farmer, the economy grew to support two products and one service—mending fences.

You are probably asking, where's the money?

Money is the blood of an economy and represents something that everyone values, such as gold or silver, that can be exchanged for any goods or services. For example, the dairy farmer might exchange a gallon of milk if the second hay farmer gave him an ounce of gold. The dairy farmer can exchange the ounce of gold some time in the future with someone else for goods and services that the dairy farmer needs.

It was originally quite difficult to get a nugget of gold or silver that weighed exactly an ounce, because nuggets are not uniform. Eventually nuggets were formed into coins that had a unified weight. Carrying lots of coins around was impractical, because they weighed too much. Coins were replaced by paper notes called bank notes issued by banks. A person could take the bank note to the bank to redeem the face value of the note in gold or silver. Bank notes were replaced by paper currency issued by the government.

Fundamentals of the primitive barter economy are still found in today's complex economy, which is a blend of a free, monopolistic, duopolistic, and oligopolistic markets.

FREE MARKET

A free market is an economy where there are few barriers to supply products or services to meet market demand. For example, in a free market economy, anyone can manufacture disposable gloves and attempt to sell them to healthcare facilities. If the disposable gloves fail to meet the demand of healthcare facilities, then the company no longer manufactures them. If the disposable gloves meet the demand, then the company makes money and continues to manufacture them.

Demand for a product or service determines if the product or service is provided. If hospitals require disposable gloves, then someone will manufacture disposable gloves. Equilibrium exists between supply and demand. If demand is low and there are too many suppliers, the profit incentive is lost and suppliers will leave the market. If demand is high and there are too few suppliers, then the profit incentive will cause suppliers to enter the market.

The equilibrium between supply and demand in a free-market economy establishes a fair price for goods and services. If demand is high and there are few suppliers, other suppliers will see this as an opportunity to make a profit by increasing the supply. As supply increases, prices usually decrease because consumers can shop around for the best price.

MONOPOLY MARKET

Although a free market is ideal, economies are typically not truly free. There are influences that control a market economy. One of these is a monopoly. A monopoly occurs when there is no competition. Consumers don't have a choice of suppliers—or suppliers have one consumer. Therefore, supply and demand is imbalanced in favor of the supplier or consumer who holds the monopoly.

A monopoly occurs naturally where there is significant savings or incentive to be realized by one person or company providing the service. In some parts of the country, a hospital may have a monopoly because it is too costly for another healthcare facility to open in the area. A monopoly also occurs through legislation, such as in the pharmaceutical industry, where a pharmaceutical company is given a monopoly to sell a drug for a particular time period as an incentive to invest in research and development of new drugs.

DUOPOLY MARKET

A duopoly market is where two dominant suppliers control a product or service, such as is seen in a small rural community that has only two physicians. There is little competition in a duopoly market economy, because consumers are limited in their choice of suppliers. It is not at all uncommon for both suppliers to offer similar or the same price.

OLIGOPOLY MARKET

An oligopoly market is the most common, and is created where several large suppliers provide an identical or similar product. While anyone can enter the market, the significant cost to participate in the market is sufficient to discourage most people. Suppliers in an oligopoly market price their product practically at the same price. Consumers have little choice when purchasing the product. For example, there might three firms that manufacture motorized wheelchairs. They form an oligopoly, because the cost of entering the market outstrips potential profit.

The Development of the Unique U.S. Healthcare Economy

The healthcare system in the United States is a third-party payer system where an insurer or government agency pays a healthcare provider for medical care given to a patient. This gives many patients the illusion that medical care is relatively free, because the patient who has medical insurance or who is covered by a government program doesn't see a medical bill. The healthcare provider sends the bill directly to the insurer.

This wasn't always the case. Before 1920, medical science was relatively primitive by today's standards. Most treatments, including surgery, were performed at home using only basic medication. As a result, medical expenses were very low and paid directly by the patient. Families had "sick insurance" in those days to cover lost wages that resulted from the breadwinner being out of work due to illness or injury. Insurance companies didn't offer health insurance, because there was a relatively low demand for the product, and there was a high risk of fraud.

New and improved medical techniques and the development of drugs to treat disease evolved after 1920, and with this came an increase in both the demand for healthcare and the cost of healthcare. Governments in Europe and the United States realized the future cost of healthcare would exceed most patients' ability to pay, and thus they set out to require mandatory medical insurance. Many European countries adopted compulsory national health insurance for everyone.

However, compulsory national health insurance didn't pass in the United States for two reasons. First, the need wasn't immediate. Medical expenses remained very low. Second, physicians and the insurance industry lobbied the government to not make it compulsory. Physicians felt the government would limit their fees. Insurance companies saw health insurance as an unmanageable risk (only sick people would buy it), and believed that if they offered health insurance they would be prohibited from offering burial insurance—a large part of their profit.

As anticipated, by 1930, the demand for new medical procedures grew and medical costs increased. It was also during this period that more people became sick, because an increasing percentage of the population moved from farms to cities where families lived closer together, allowing diseases to spread.

Around this same time, requirements for practicing medicine were established by the American Medical Association (AMA). Physicians were required to graduate from an accredited medical school. Medical schools toughened admission requirements and instituted stricter graduation requirements. As a result, the number of physicians dropped. Physicians who remained in medicine took advantage of this situation and increased their medical fees. Non-hospital medical costs for an average family increased from 5 percent of their income to 13 percent.

Hospitals were in a crisis. There was a high cost of hospitalization, but patients' wages were low. Hospitals treated patients, but had to wait for the patient to pay the bill. Many bills went unpaid because patients couldn't afford to pay them.

THE FIRST MEDICAL INSURANCE

In 1929, Baylor University Hospital joined together with a group of Dallas teachers to create a prepaid hospital service. For $6 a month, members of the group could receive 21 days of hospitalization at Baylor University Hospital per year. As a result, the hospital received a steady flow of income before giving treatment. Teachers knew that the cost of a hospital stay would be covered by the $6 per month prepaid premium. In 1931, the American Hospital Association became involved with the concept of health insurance, and encouraged its expansion into other hospitals. By 1946, the insurance component became independent from the hospitals, calling itself the Blue Cross Commission and, by 1972, simply Blue Cross.

Blue Cross subscribers still had to pay physicians' bills, however. Physicians didn't want to join a prepaid medical service, because they wanted the freedom to charge patients based on each patient's capability to pay, rather than charge a fixed fee. Eventually, the AMA developed guidelines for prepaid medical service for physicians. This eventually developed into Blue Shield, around 1948. Under this system, the physician's bill is sent to the patient. The patient submits the bill to Blue Shield, which reimburses the patient a fixed amount, usually less than the full amount of the bill. The patient has to pay the difference.

In the past, Blue Cross was exempt from insurance regulations and was allowed to function as a tax-exempt, nonprofit organization because it was considered a service to the community in that it provided benefits to low-income individuals. Today, individual states determine whether or not to give Blue Cross this designation and exemption. In New Jersey, for example, Horizon Blue Cross/Blue Shield is a not-for-profit corporation, but it is not tax exempt.

EMPLOYER-BASED MEDICAL INSURANCE

Between 1940 and 1960, the insurance industry learned how to handle the risk of fraud in medical insurance and began offering medical coverage. However, the for-profit insurance industry used a different strategy from that of Blue Cross and Blue Shield. It targeted employers rather than individuals, and set an insurance premium according to the actual experience of illness among the employee population, rather than at a fixed rate. That is, employees in poor health paid a higher premium than their healthier counterparts, rather than the fixed-fee rate charged by Blue Cross and Blue Shield. This meant that the employer's premium reflected its employees' demand for healthcare

Employers realized that offering healthcare coverage was a way to attract employees during World War II. In 1942, Congress passed the Stabilization Act, which created wage and price control. Employers were limited in the raises that they could give to their employees. One objective of the Stabilization Act was to decrease competition among employers for employees, since there was a shortage of workers. Most workers were in the armed service. An employer couldn't offer an employee more money to change jobs, but the employer could offer health insurance. Health insurance became a benefit of working for an employer.

The United States was a manufacturing economy in the 1940s. Unions represented factory workers and negotiated terms of employment. The end of World War II brought an end to the worker shortage. By 1949 employers wanted to reduce costs by canceling medical insurance for their employees. However, unions persuaded the federal government to prevent an employer from canceling or modifying group health insurance during the term of a contract. Furthermore, the federal government ruled that medical insurance was part of an employee's wages, and therefore the employee or the union had the right to negotiate medical insurance as part of the wage package. This decision created a de facto employer-based health insurance system that is still in existence today.

1965 SOCIAL SECURITY ACT AMENDMENT

Although the emergence of employer-based healthcare insurance significantly reduced the number of U.S. citizens who did not have health insurance, a quarter of the people in the United States were not covered by health insurance in 1958. Most of these were the unemployed, sickly, and elderly. The unemployed had no money to pay for health insurance. Medical insurers set premiums very high for the sickly and elderly, because premiums were set based on experience—the more claims a person filed, the higher the insurance premium—so they too couldn't afford to pay the premium.

In 1965, Congress passed the Social Security Act amendment that created Medicare and Medicaid. Medicare is the government-sponsored health insurance, financed through payroll tax, that covers anyone 65 years or older and certain younger people with disabilities. Medicare is divided into two sections, Part A and Part B. Part A provides hospitalization insurance. Part B pays for physicians' services and reimburses the physician directly for the physician's usual, customary, and reasonable rates. The patient is exempt from any co-payment to the physician when the physician accepts the Medicare fee.

Medicaid is a federal and state run medical insurance program for patients whose income is below the state's poverty level. A state receives money from the federal government based on the state's per capita income compared with the national per capita income. The federal government specifies minimum eligibility requirements and benefits. Each state can increase them. Therefore, eligibility requirements and benefits for Medicaid differ from state to state.

The number of Medicare subscribers has grown to a level that stretches the government's capability of funding the program. In 1983, the government changed how healthcare providers are reimbursed. Instead of reimbursing at the usual and customary rates, the government Medicare and Medicaid programs now reimburse based on a prospective fixed-fee schedule. This has resulted in many providers not accepting Medicare or Medicaid as a form of payment.

The Business of Healthcare

Healthcare is a business of making money by caring for the sick. However, the way the healthcare industry does business is radically changing. Employers shop around to find the most cost-effective medical insurer and, to compete, medical insurers devise ways to make their medical costs predictable by creating networks of healthcare providers. Healthcare providers who join the networks agree on negotiated fees.

Medical insurers use economic incentives to encourage patients to use in-network healthcare providers. Patients who use in-network healthcare providers have little or no out-of-pocket expense. The patient may have to pay up to 30 percent of the healthcare provider's fee out-of-pocket if the healthcare provider is out-of-network. Likewise, medical insurers may pay 100 percent for generic medications and 70 percent for brand-name medications.

Medical insurers are trying to educate the consumer to the financial aspect of a healthcare transaction. The first sign of change occurred when some healthcare providers resisted joining a network. They changed their mind once patients left and went to in-network healthcare providers.

A battle is brewing between third-party payers and healthcare providers, as third-party payers try to control increasing medical costs, and healthcare providers try to preserve income, profit, and independence. Consumers are sometimes caught in the middle, resulting in consumers who want the highest quality of care without having to pay more for it. In many situations, health insurers require that healthcare providers receive approval from the insurer's medical staff before performing tests or giving treatment. Many healthcare providers see this as infringing on their medical judgment. Up to now, rarely did a healthcare provider have to justify their medical decisions. Some patients feel that medical insurers are looking to increase their profit by forcing the healthcare provider to give the patient second-class medical care.

Some healthcare providers have raised their published fees to unrealistic levels, anticipating that their published fees are the starting point for negotiating fees they'll receive from a third-party payer. If they want to receive $100 per visit, they publish a fee of $200 so that they can negotiate away 50 percent of their fee without incurring a loss of income. Unfortunately, patients who are not covered by medical insurance find themselves paying fees that are unrealistically high.

The federal government's recent Medicare Prescription Drug Plan is the latest attempt to reel in drug costs. The plan grants medical insurers the right to offer approved drug coverage to Medicare patients. Medicare patients are required to purchase drug coverage from an approved medical insurer. In theory, medical insurance plans woo Medicare patients by offering a competitive premium that reflects the cost-effective arrangements the medical insurers have made with drug suppliers. Medicare patients are required to pay for medication that is not covered by the medical insurer. The idea is to have Medicare patients ask their physicians if they can switch to less costly drugs that the medical insurer covers. However, the Prescription Drug Plan became a bureaucratic nightmare, leaving many Medicare patients confused, and some paying more for medication then they did under previous drug plans.

Economists understand that moving to a healthcare system in which consumers pay more of the bill is the most efficient way to control spiraling healthcare costs.

However, there is a catch-22: Patients cannot afford to pay the price for healthcare and want to have the low-cost out-of-pocket expense that many have now. Healthcare providers lose income by lowering fees, accepting preset Diagnosis-Related Group fees, and paying the generally rising administrative costs of doing business.

Nurse Managers and Healthcare Economics

Nurse Managers are on the front lines in the effort to provide cost-effective patient care, because they must work with available funds to acquire all resources necessary to provide medical care to patients.

As a Nurse Manager, your unit is a business unit that has income and expenses. Income comes from payment for medical care given to patients. Expenses are the cost of providing the medical care. As in any business, the healthcare facility is expected to bring in more income than is expended on patient care. In a not-for-profit facility, excess income can be used to expand the facility's service to the community. In a for-profit facility, excess income is returned to investors.

Healthcare facilities are governed, to some extent, by the law of supply and demand. That is, the healthcare facility fulfills the community's healthcare demands. It is the responsibility of senior healthcare administrators to identify current and future demands for healthcare, and then create a healthcare organization that can meet the demand. Healthcare is not an exact free-market, supply-and-demand business, because the demand is very elastic. That is, patients often require care immediately and they cannot plan its purchase. It is not a consumer item, such as milk, that can be put off or just not purchased. The demand is immediate and the fluctuation of the demand is hard to anticipate. Will we have a serious flu this year? Maybe an explosion will send you many burn patients? Your orthopedic unit may gear up for winter fractures that never materialize but flu and pneumonia patients clog the medical unit. This is why cross-training nurses is so important.

Careful analysis of the community identifies trends that could forecast new healthcare demands. For example, a community with an aging population usually has an increasing need for orthopedic care. Therefore, senior healthcare administrators can assume there will be an increasing demand for orthopedic care. Senior healthcare administrators project annual demand, operating expenses, and income for a five-to-ten-year period. The demand is represented as the number of patients cared for by the unit. Operating expense is the ongoing cost of caring for these patients. Income reflects the fees received for giving this care. If the income sufficiently exceeds expenditures, then the healthcare facility's trustees can invest in building and furnishing a new orthopedic unit. The one-time cost to build and furnish a new unit is referred to as a *capital expenditure*.

The hospital administration establishes three annual goals by analyzing data from previous months and years, for the Nurse Manager of the new orthopedic unit:

- Target number of patients
- Planned expenditures
- Expected income

Senior healthcare administrators are simply saying to the Nurse Manager, "We expect you to spend X amount of money to care for Y number of patients and return to us Z income." If the Nurse Manager meets those goals, then the unit generates the anticipated financial goals. And, if other units meet their similar goals, the healthcare facility earns excess money.

Planned expenditures are incorporated into the unit's budget. The budget is a financial guide for operating the unit. The budget is divided into categories of expenditures, such as staffing, supplies, telephone, and other costs incurred when operating the unit. Each category is assigned an amount of money that can be expended. The Nurse Manager is responsible to keep expenditures within amounts specified in the budget, or to provide a reason for any wide variance.

The annual budget is broken down into 12-month intervals that represent milestones for the year. At the end of each month, senior healthcare administrators review expenditures and income for that period to determine if the unit is operating on target and will reach its annual goals. Budgets are set for the fiscal year and no adjustments are made, but a variance analysis is ongoing to determine what parts of the budget are, or are not, on target.

A budget may be off target because of inaccuracies in one or more of the three projected goals. That is, there may be fewer or more patients than predicted; costs may be lower or higher than anticipated; or income may be lower or higher than projected. The Nurse Manager is not expected to directly influence the number of patients admitted to the unit. Indirectly, the Nurse Manager influences the number of patients by providing quality patient care that encourages physicians to use the hospital. Patients are sometimes limited to which healthcare facilities they can use, based on the insurance they possess. The Nurse Manager is also not directly responsible for the reimbursement dollars, because the Nurse Manager does not negotiate fees with HMOs, etc. Shortening the length of stay, however, will increase Diagnosis-Related Group (DRG) reimbursement. The area with the greatest Nurse Manager impact is the costs related to salaries and supplies.

ECONOMIC CHALLENGES

There are far-reaching influences beyond the Nurse Manager's control that make it challenging to meet goals established by senior hospital administrators. Goals are set based on assumptions that certain things are going to happen and happen on time. However, forecasting healthcare demand is different from forecasting demand in other businesses. In a gas station's business, there are weekly repeat customers who purchase approximately the same amount of fuel each week. Healthcare is not that predicable, because repeat business isn't as likely. For example, a patient who has knee surgery today is unlikely to return anytime in the future once the knee heals.

Healthcare goals represent the likelihood that a certain number of patients in a community will require a specific type of care at a particular hospital. In budget preparation, a hospital looks at census numbers for all the units in the hospital and projects the coming year's numbers based on the previous year's patient days. If orthopedics had 14,000 patient days the previous year, the hospital may project 15,000 this year, if two new orthopedic surgeons have been added to the medical staff. In most hospitals, the patient-day budget will be spread evenly over 12 months, even though the administrator and the Nurse Manager know that the winter months will bring more patient days due to falls in inclement weather. In times like this, the Nurse Manager temporarily adjusts staffing levels by temporarily hiring agency nurses or using more overtime and per diem help.

Income goals are also difficult to achieve sometimes for a number of reasons. Some patients don't pay, which results in uncompensated care. Prior to the spiraling healthcare costs, senior healthcare administrators built this loss into the healthcare facility's fee structure. That is, those who could pay, paid an additional amount to cover the cost of uncompensated care. This is referred to as *cost shift*.

However, today, third-party payers negotiate fees that exclude a cost shift. As a result, the healthcare facility is unable to generate excess income from paying patients to absorb the cost of uncompensated care. Some state governments provide supplemental payments to healthcare facilities for uncompensated care. Healthcare facilities are at risk of shutting down if they have too many uncompensated care patients.

The payer mix can be another obstacle for not meeting income goals. The payer mix refers to the mix of how patients pay for healthcare. The mix includes full-price payers (patient pays the published fee), managed care organizations (which pay a negotiated fee), and insurers (discounted fee). Patients who have medical coverage are referred to as *covered lives*, and the payments they make to an insurer are called *premiums*. Although senior healthcare administrators project a certain payer mix, the actual payer mix may differ significantly from the projection. Furthermore, their projections anticipate a specific negotiated or discounted fee with third-party payers that may not be realized.

The federal government is the largest third-party payer, accounting for 50 percent of the funds spent on healthcare. Federal government healthcare programs include Medicare, Medicaid, the Federal Employee Health Benefits Program (FEHB), and the Civilian Health and Medical Program of the Uniform Services (CHAMPUS). The federal government pays healthcare providers based on a fee schedule using the DRGs. Each DRG is assigned a fee. The government can change the fee, resulting in an unanticipated drop in income for the healthcare facility.

Expenses are also difficult to predict. Although efforts are made to ensure that the budget reflects an estimate of future costs, actual costs can increase during the year. This requires the Nurse Manager to find cost-effective alternatives to the budgeted item, or to transfer funds earmarked for other purchases.

Healthcare is a highly regulated industry. Forecasts of expenses and income are partly based on the impact that existing and pending regulations have on the unit's operations. Those regulations, however, can change, resulting in possibly higher cost and/or lower income.

MANAGING ECONOMICS OF YOUR UNIT

The Nurse Manager runs a healthcare business. The Nurse Manager's unit either makes or loses money for the healthcare facility. The initial step in running a successful unit is to make sure the unit's annual goals are realistic. Although senior healthcare administrators set the goals for the number of patients, expenses, and income for the unit, the Nurse Manager should carefully compare these goals to the unit's current operation.

A Nurse Manager must ask the senior healthcare administrators for the rationale for goals that seem unrealistic based upon their experience operating the unit. Let's say that for the past two years, the unit has cared for 50 patients a month and the unit is forecast to care for 100 patients per month. The forecast is not consistent with the current trend, therefore the Nurse Manager must determine the reason for the expected jump before agreeing to the goal.

As a Nurse Manager, don't be afraid to challenge the basis for the forecast, because your position might expose you to information that isn't obvious to senior healthcare administrators. Likewise, senior healthcare administrators may have information that is not available to you. For example, they may know that a neighboring hospital is closing the counterpart to your unit, which will increase the number of patients coming to your unit. Also, some units that do not make money on their own, feed other services that do. For example, some hospitals may maintain a home health agency in order to be a full-service facility. This gives the facility an advantage vis-à-vis other facilities that do not have home care, physical therapy, etc. Physicians and their patients prefer to have all their healthcare needs met within one healthcare system.

The expense goal for your unit reflects the range of patient classifications that senior healthcare administrators expect for the coming year. As you'll recall from Chapter 3, healthcare facilities use a patient classification system to determine the number of hours per day that each type of healthcare provider spends caring for a patient that has a specific DRG. This translates into an hourly cost to care for the patient and becomes an expense to the unit. A Nurse Manager must make sure that the range of patient classifications used to set the goal reflects the current trend; otherwise, actual expenses might be higher than expected.

Summary

No one is self-sufficient. Things we cannot do for ourselves create a demand for others to do it for us, which is the basis for an economy. An economy is a system of production, distribution, and consumption. The value of a product or service is influenced by demand of and availability of a product or service.

Today's economy, in the United States, is a blend of free, monopolistic, duopolistic, and oligopolistic market economies. A free market is an economy where there are few barriers to supply products or services to meet market demand. A monopoly occurs when there is no competition. A duopoly market is where two dominant suppliers control a product or service. An oligopoly market is the most common, and is created when several large suppliers provide an identical or similar product.

Until the 1930s, the healthcare industry adhered to supply and demand of the marketplace. Consumers directly purchased healthcare, forcing healthcare providers to make healthcare affordable. An economic imbalance occurred about this time, when medical science matured and the population increasingly shifted from farms to cities. Families lived close together, allowing diseases to spread. Demand for healthcare increased. Hospitals fulfilled this demand. However, the cost of hospitalization outstripped the consumer's ability to pay.

In the 1930s, Baylor University Hospital joined together with a group of Dallas teachers to create a prepaid hospital service. This evolved into Blue Cross. A similar service, called Blue Shield, was created by the AMA for prepaid medical service for physicians.

Another economic shift affecting the healthcare industry occurred during World War II. In 1942, Congress passed the Stabilization Act, which created wage and price controls. Employers were limited in the raises that they could give to their employees, which made it difficult to attract new employees. An employer couldn't offer an employee more money to change jobs, but could offer health insurance. This was the start of today's employer-based health insurance system.

The employee consumed healthcare delivered by the healthcare provider, but paid nothing. The medical insurer paid the healthcare provider and then passed along the cost to the employer. The employer didn't pass along costs to the employee, because healthcare cost was a benefit. This disrupted normal economic forces that kept supply, demand, and pricing in balance.

Quiz

1. A system of production, distribution, and consumption is called:

    (a) An exchange system

    (b) Economy

    (c) A business

    (d) None of the above

2. An economy where there are few barriers to supply products or services to meet market demand is called a:

    (a) Free market

    (b) Monopoly

    (c) Duopoly

    (d) None of the above

3. What is the best method to control prices?

    (a) Third-party payer

    (b) Government regulations

    (c) Consumer-paid services

    (d) None of the above

4. The federal government developed the employer-based health insurance system.

    (a) True

    (b) False

5. The healthcare industry is not affected by the laws of supply and demand.

    (a) True

    (b) False

6. Covered lives means:

    (a) Patients who have medical coverage

    (b) Patients whose medical coverage has lapsed

    (c) Patients whose medical coverage is soon to lapse

    (d) None of the above

7. The federal government is the largest third-party payer.

    (a) True

    (b) False

8. What is it called when patients who pay healthcare directly or through a third-party payer also pay for patients who are unable to pay for healthcare?

    (a) Cost shift

    (b) Charity care

    (c) Reallocation of revenue

    (d) None of the above

9. Care given to a patient who is unable to pay for healthcare is called uncompensated care.

    (a) True

    (b) False

10. The payer mix is a combination of patients who directly pay for healthcare and various third-party payers.

    (a) True

    (b) False

CHAPTER 9

Budget Planning and Financial Management

Mention the title Nurse Manager and the last thing that comes to mind is someone who is responsible for running a money-making unit for the healthcare facility. Yet, it's true. A Nurse Manager is more a business manager at times than someone who manages a patient's healthcare.

In essence, the Nurse Manager runs both the healthcare side and the business side of a unit, and must be as comfortable with finance and budgets as with patient assessments and treatments.

In this chapter you build on your knowledge of healthcare economics learned in Chapter 8, and explore the financial aspects of operating a unit. You'll learn about budgets, cash flows, trends, and other tools a Nurse Manager uses to manage the business side of a unit. Baccalaureate nursing programs usually have management and leadership nursing courses, but many do not cover these issues sufficiently to prepare you to manage a unit without some additional instruction. There is a move among nursing organizations and accrediting bodies of schools of nursing for nurses to possess a master's degree for entry-level management.

Financial Structure of a Healthcare Facility

As you learned in the previous chapter, a healthcare facility is a business that supplies healthcare in return for money. Fees charged for healthcare should cover expenses and return an amount above the cost. This is referred to as *excess earnings*. In a not-for-profit healthcare facility, excess earnings are used to expand healthcare services to the community. In a for-profit institution, excess earnings are considered profit and returned to investors.

A healthcare facility is financially organized into accounts much like your banking account, except these accounts are maintained by the accounting department rather than a bank. Accounts are generally grouped into revenue centers and cost centers. A *revenue center* is an account used to record income. So when a patient's insurer pays for ER care, the money is recorded in the ER's revenue center. The actual money is deposited in a bank. Think of a revenue center as a score card on which the ER is given credit for bringing in the money. A *cost center* is an account used to record expenses. So when the ER nurse gives a patient an injection of epinephrine, the cost of the medication, syringe, needle, the nurse, and other things necessary to give the injection are recorded as expenses in the ER's cost center.

Each department of the healthcare facility is assigned a cost center, and some departments are also assigned a revenue center depending on the nature of the department:

- ER, surgery, and other similar units that generate revenue are assigned both a revenue center and a cost center to record revenue they receive and expenses they incur.

- Security, building maintenance, and other departments that don't generate revenue are assigned a cost center to record expenses they incur. These costs are usually considered overhead and shared among all the revenue-generating departments.

- Pharmacy, central supply, and similar departments are assigned both a revenue center and a cost center in some healthcare facilities, because they operate as an internal business to control the healthcare facility's expenses.

Let's say that the ER requires a supply of epinephrine. The ER "buys" epinephrine from the pharmacy. Think of the pharmacy as "selling" epinephrine to the ER. The "fee" the pharmacy charges is recorded in the ER's cost center—and the fee is also recorded in the pharmacy's revenue center. This is referred to as a charge-back. In this way, the pharmacy will purchase only medications that can be charged back to a revenue-generating department. The ER passes along the cost of the epinephrine in the fee charged to the patient.

Each revenue-generating department is expected to produce excess earnings. Excess earnings are determined by subtracting the sum of the cost center from the sum of the revenue center. A positive number is excess earnings. A negative number is a deficit, which means the department spent more on expenses that it received in revenue.

BUDGETS

A *budget* is an accounting document that contains the healthcare facility's revenue and expense goals for the year. It contains expected revenue and expenses. In contrast, a revenue center and cost center contain actual revenue and expenses.

A healthcare facility's budget is very similar to your household budget. Your household budget sets your income and expense goals. However, the healthcare facility's budget is more complex than your household budget, because it must adhere to strict accounting regulations.

The healthcare facility's overall budget is divided into budgets for each department. Revenue-generating departments have both revenue and expense budgets. The revenue budget specifies the amount of revenue the healthcare facility administrators expect the department to generate. The expense budget contains expected expenses for the department. Non-revenue-generating departments have only an expense budget.

CHARTS OF ACCOUNTS

As mentioned earlier in this chapter, a healthcare facility is financially organized into accounts grouped as either revenue centers or cost centers. Revenue centers and cost centers are subdivided into a standard series of subaccounts referred to as the *chart of accounts*. Each subaccount is further divided into additional subaccounts. Each subaccount is assigned a unique number and description. Although you don't have to memorize the chart of accounts as a Nurse Manager, you do need to be familiar with the chart of accounts that pertains to your unit. Your accounting department provides you with this information. The chart of accounts is used in budgets and in revenue and cost centers.

Alongside each account is a dollar amount. In a budget, the dollar amount is the expected amount of revenue or costs for the year. In revenue and cost centers, this is the actual income and expense. For example, there might be a subaccount called per diem registered nurses, which contains the cost for per diem registered nurses for the unit. The budget's per diem registered nurses item reflects the expected expense, while the same subaccount in the cost center reflects the actual expense.

BUDGET PLANNING

Budget planning is a collaborative process involving all levels of management to devise the healthcare facility's budget. This process begins with a review of the healthcare facility's strategic plan. A *strategic plan* defines long-term goals for the facility to supply future healthcare needs to the community.

Let's say the healthcare facility serves an aging community. The strategic plan specifies that the facility will have a strong presence in orthopedics, cardiac surgery, ER, and diabetic and cancer care. The strategic plan estimates the number of patients per unit for each of the five years covered by the plan.

After reviewing the strategic plan, the next step is to determine the current healthcare needs of the community. Think of the strategic plan as a best estimate. The healthcare needs over the last month and in recent years are a good reflection of current needs.

Next, determine if there is any revolutionary change that might be foreseen in technology, medicine, in the community, or in the way healthcare facilities are re-imbursed. The business of healthcare is usually evolutionary, where subtle changes occur each year. Sometimes there are radical changes that disrupt the normal way healthcare is provided to patients. For example, recent technology changes enable x-rays and other tests to be evaluated nearly instantaneously by qualified, lower-paid professionals in offshore countries such as India. This is a revolutionary change that can have a material impact on the healthcare facility's business. The healthcare facility's cost for x-rays is likely to be substantially higher than that of other health-care facilities who take advantage of offshoring.

Based on review of the strategic plan, recent experience, and determination of the impact of changes in the industry, senior managers decide which services the healthcare facility will offer patients next year. Typically, these are the same services offered in the previous year. However, sometimes senior managers decide to offer a new service, or stop offering an existing service, depending on demand.

After senior managers set the high-level objectives for the healthcare facility, unit managers and middle managers devise budgets for each revenue-generating unit. They estimate how many and what kind of diagnoses will be treated by each revenue-generating unit, based on trends in patient care. Once the number of each kind of diagnosis is estimated, managers estimate the cost to treat each diagnosis by itemizing everything required to care for a patient with that diagnosis, and the length of stay to deliver that care.

The pharmacy, HR, central supply, and other departments within the healthcare facility provide cost estimates for items they supply to the unit. Their estimates reflect expected price changes in the market for the year. For example, HR provides an estimate for the hourly cost of an RN. A good estimate is within plus or minus 10 percent of the actual cost.

CALENDAR VS. FISCAL YEAR

A budget is a forecast of revenue and expenses for a 12-month period, after which the revenue and expenses are reconciled and the healthcare facility determines if there are excess earnings.

Some healthcare facilities use a calendar-year budget, which runs from January through December, whereas others use a fiscal-year budget, which covers any 12 consecutive months. Your household budget uses a calendar year, primarily because it is convenient and you don't have to follow strict accounting regulations. However, a business such as a healthcare facility must adhere to the accounting principle that states income and related expenses should be aligned as closely as possible in the budget year.

Some businesses, such as retailers, might experience higher income toward the end of a year, and related expenses at the beginning of the next year. For example, December is a high-income period for retailers, and January is a high-expense period as customers return items purchased in December. A calendar-year budget isn't appropriate for retailers, because high expenses in January relate to December's income. Therefore, the business uses a fiscal year budget that may be from June through May.

Your healthcare facility's budget planning coincides with its budget year.

CAPITAL BUDGET VS. OPERATING BUDGET

There are two categories of expenses that are estimated in a budget: operational expenses and capital expenses. An *operational expense* is incurred to operate the healthcare facility for the year and includes such items as salaries, electricity, and medicines. A *capital expense* is for a long-term asset such as digital patient monitors, renovating a unit to serve as a critical care unit (CCU), and constructing a new wing for orthopedic surgery.

Think of an operational expense as an item that lasts a year or less, and a capital expense as an item that lasts more than a year. Some healthcare facilities use a dollar amount to differentiate between an operational and a capital expense. For example, items that cost $2,000 or less are consider an operational expense, even though the item may last more than a year. Items over $2,000 are considered a capital expense. This is similar to you buying a $500 television, which you'll probably pay for using income you earn this year, versus buying a new car that you pay for over several years. Both the television and car will last more than a year, but it makes fiscal sense to pay for the television within the year. Also, issues of depreciation affect capital and operating budgets. By the time a facility receives new computers, there is already something bigger and better available, and some equipment is quickly obsolete.

PRODUCT-LINE BUDGET

A healthcare facility may organize some budgets along product lines, which are referred to as *product-line budgets*. A product line is a group of services provided by different units in the healthcare facility to care for a patient who has a specific diagnosis. The product line generates income just as a revenue-producing unit does.

As an example, suppose that treatment for colon cancer is a product line. This treatment involves care given by several revenue-producing units, such as radiology, surgery, and the medical-surgical unit. The product line is assigned a revenue center and a cost center, along with a product-line budget. For budgetary purpose, the radiology department charges the product line a fee for patient care, as do the surgical and medical-surgery units.

A product-line budget enables senior management to associate revenues and expenses for the treatment of a common diagnosis for the healthcare facility.

SPECIAL-PURPOSE BUDGETS

An operational budget is used to manage recurring healthcare services, such as treating a colon cancer patient. However, there are times when the healthcare facility offers a one-time service to meet the immediate needs of the community, such as screening for skin cancer or cholesterol. These one-time activities are not included in the operational budget. Instead, a special-purpose budget is created, along with a corresponding cost center, or maybe a revenue center depending on the nature of the activity. Special-purpose budgets are created by the manager who is responsible for the activity.

CASH FLOW

A healthcare facility financially operates similar to your household. You and the healthcare facility incur expenses daily, resulting in bills, and receive income used to pay those bills. In a perfect world, income is received before bills come due. In the real world, sometimes a bill comes due before there is cash to pay the bill. The flow of cash into and out of a healthcare facility is referred to as *cash flow*.

The goal of every healthcare facility is to maintain a positive cash flow. A positive cash flow means that at any point in time there is sufficient cash coming from income to cover expenses. There are periods during the year when the healthcare facility might experience a negative cash flow. A *negative cash flow* occurs when there is insufficient cash coming from income to pay the bills. When this happens, the treasurer of the healthcare facility borrows money from a financial institution to cover bills that are coming due. The loan is referred to as *revolving credit*, which is similar to using your credit card. Revolving credit is a short-term loan that is usually completely paid within the month. Borrowing funds is expensive, because financial institutions charge premium interest rates for short-term loans.

Healthcare is unlike the supermarket where the customer pays cash for purchases before leaving the store. In healthcare, a detailed invoice and support documents must be submitted to a third-party payer. Ninety days or more pass before payment is received. Complicating matters, each third-party payer has rules that must be followed before they pay an invoice. For example, some third-party payers require prior approval for expensive treatment, or cover only a percentage of cost for brand-name medications. The patient must directly pay the costs not covered by the third-party payer. As a result, there can be unexpected delays in receiving payment from a third-party payer. Furthermore, the healthcare facility must invoice the patient for any charges not paid by the third-party payer. In some cases, patients are unable to pay these charges.

Healthcare facilities attempt to minimize the delay in payment by developing a close working relationship with major third-party payers. Just as in any business, the timing of the receipt of reimbursement, which is tracked by accounts receivable, is of the utmost importance. To assist in this matter the healthcare organization takes certain steps.

Before the patient is admitted:

- The healthcare facility staff resolves any potential conflicts regarding the proposed treatment with the third-party payer.

- Adjustments in treatment might be made to ensure that care will be covered by the third-party payer.

- The patient is informed of any expenses that they will be responsible for paying.

While the patient is in the hospital:

- Nurse managers carefully monitor patient care to ensure that the patient receives the approved treatment.

- Expenses are recorded and supporting documents required by the third-party payer are prepared.

- If the physician wants to deviate from the approve treatment plan, the condition of the patient and alternate treatment plans are discussed with the third-party payer.

After the patient is discharged:

- The invoice is immediately prepared and sent to the third-party payer.

- The patient is invoiced immediately for charges not covered by the third-party payer.

- The accounts receivable department follows up with both the third-party payer and the patient for payment.

By focusing on the needs of the third-party payer before the patient is admitted and while the patient is being treated, the healthcare facility reduces the likelihood that payment for an invoice will be unexpectedly delayed from a third-party payer.

The Nurse Manager's Fiscal Responsibilities

The Nurse Manager is responsible for the care of patients on their unit, as well as the unit's financial operation. Several months prior to the start of a new fiscal year, the Nurse Manager is expected to develop a budget for the unit, which is then reviewed, adjusted, and eventually accepted by senior management.

Once the budget is approved, the Nurse Manager provides patient care within the constraints of the budget. At the end of the fiscal year, senior management expects the unit's actual revenue and expenses to be within ±10 percent of the budget. Realistically, situations arise where revenues and expenses can be beyond this range, such as an unexpected increase in admissions, or a new treatment enabling patients to be treated as on an outpatient basis, resulting in lower than expected admissions. The Nurse Manager must provide a rationale for any variation where the actual amount is ±10 percent of the budgeted amount. Hospitals may set different percentage figures, but rationales are always required.

BUDGET PLANNING

Senior managers look to the Nurse Manager of the unit to prepare the unit's budget, because the Nurse Manager is most knowledgeable about the unit's operation. In essence, the senior manager asks the Nurse Manager the following questions:

- What types of diagnoses does your unit expect to treat next year?
- How many patients do you expect for each diagnosis?
- What is the minimum time to treat each diagnosis?
- What is the average time to treat each diagnosis?
- What is the maximum time to treat each diagnosis?
- How much will it cost to treat these patients?
- What time of the year do you expect patients to be admitted to your unit? (More admissions in the winter during flu season?)
- What new equipment do you need to treat these patients?
- What are your staffing requirements for next year?

The Nurse Manager's responses become the basis for the unit's budget; therefore, the Nurse Manager must be prepared to present senior management with a rationale for each response. A rationale must be based on the Nurse Manager's analysis of empirical data related to the unit. *Empirical data* is data gathered from past experience of unit activities.

For example, patient records provide empirical data to the finance department to learn the type and number of diagnoses the Nurse Manager's staff treated last year, as well as the length of treatment. The admission records show when patients were treated during the year. The healthcare facility's accounting department can probably identify the revenue generated by and the cost to treat each diagnosis. And the HR department can provide information about actual staffing for the unit during the past year. The Nurse Manager's current staff is a good source to determine what new equipment is needed.

Empirical data tells the Nurse Manager what happened in the past year, but doesn't tell what is going to happen next year. The Nurse Manager shouldn't assume that what happened in the previous, or current year, will also happen next year, because changes might occur.

After gathering and analyzing empirical data, the Nurse Manager must review the healthcare facility's strategic plan to determine if the strategic plan calls for modifications in the unit's operation. For example, suppose that the strategic plan calls for a material increase in colon cancer treatment to coincide with the age of the community. The Nurse Manager must consider the impact of this increase on the unit when developing next year's budget.

Keep in mind that the strategic plan is a long-term estimate of what will happen in the future. As a rule, there is a greater error factor in long-term estimates than in short-term estimates. Therefore, the Nurse Manager must compare long-term estimates in the strategic plan with the unit's empirical data, and determine whether estimates in the strategic plan are sensible. Some situations are impossible to predict. To help alleviate some uncertainty, healthcare communities divide responsibilities in case of certain major catastrophic events. For example, hospital A will be the major receiving center for radioactive emergencies, while hospital B will be the primary receiving center for major trauma. This requires healthcare agencies to work together and develop a strategic plan. This is also done within a hospital: unit A will receive the overflow for trauma, unit B will receive overflow for biological agents. This requires additional cross-training for nursing staff.

If estimates in the strategic plan are accurate, those estimates should closely correlate with the unit's empirical data. If they differ substantially, then the Nurse Manager bases the unit's budget on the empirical data and notes the discrepancy as the rationale for disregarding estimates in the strategic plan.

The next step in preparing the budget is to determine if there are any factors that might disrupt the normal operations of the unit. These factors can range from operational distractions, such as new construction near the unit, to technological improvements that change treatment given to patients. The unit's budget should reflect all disruptive factors.

Cost estimates are the last item that the Nurse Manager needs to gather before assembling the budget. Departments that provide supplies to the unit are the best source for estimating next year's costs.

The moment of truth arrives when the Nurse Manager must set the budget for the next fiscal year. The Nurse Manager analyzes all the information gathered, and then uses his or her best judgment to create the budget.

The budget document itself is usually provided by the healthcare facility's financial department along with guidelines for submitting the budget for approval. The Nurse Manager should expect a meeting with senior management to discuss the proposed budget. During the meeting, senior managers will question any item that materially varies from the current year's budget. The Nurse Manager is expected to explain the rationale for the variance.

Once approved, the Nurse Manager is expected to adhere to the budget when managing the unit.

ZERO-BASED BUDGETS

Some healthcare facilities have adopted a zero-based budget process in an effort to reduce unnecessary expenses throughout the organization. A *zero-based budget* means that the Nurse Manager builds a budget as if it were the first budget for the unit. It is as if senior management gave the Nurse Manager a blank slate and said create a budget for a unit that will care for a specific number of patients who have specified types of diagnoses.

Zero-based budgeting requires the Nurse Manager to justify every item in the budget. This differs from traditional budgeting methods that assume next year's budget will be the same as this year's budget, perhaps adjusted for inflation. For example, in traditional budgeting, the Nurse Manager might say, "We'll need ten RNs next year, because that's the number we have this year." In zero-based budgeting, the Nurse Manager might say, "Each day we expect five colon cancer patients, three gallbladder patients, and seven patients with diabetes-related complications. Based on our experience, an RN is required to spend 3 hours each day with a colon cancer patient, 2 hours with a gallbladder patient, and 4 hours with a patient who has complications from diabetes. As a result, 15 hours per day will be spent caring for five colon cancer patients, six hours for three gallbladder patients, and 42 hours for diabetic patients. This means that, in total, the unit requires 63 hours per day of care from an RN—or about eight RNs."

In theory, zero-based budgeting is expected to reduce unnecessary expenses by matching expenses to the current level of demand—not last year's level of demand.

In practice, zero-based budgeting is time consuming and requires gathering additional information about anticipated demand for service that may not be readily available within the deadline for preparing the budget. As a result, some Nurse Managers use the current year's budget as the basis for the zero-based budget.

CAPITAL BUDGET

The Nurse Manager is responsible to propose capital expenditures for the unit to senior managers. A *capital expenditure* is the acquisition of a long-term asset, such as digital patient monitors and computers for the nurses' station. The proposal must be focused on fulfilling a need of the unit that relates to making or saving money for the healthcare facility. Senior management is willing to invest substantial sums in a capital expense, if in the long run it will make or save more money than it cost.

Let's say that the healthcare facility's insurance premiums have dramatically increased because of an increase in less-than-adequate patient monitoring on the unit. The nurse-to-patient ratio is high, resulting in fewer nurses' visits to monitor patients. Digital patient monitors are a capital expenditure that restores adequate patient monitoring by enabling every patient to be monitored at the nurses' station.

The proposal should also show how the capital expenditure will make or save money. For example, the Nurse Manager might state that digital patient monitors will lower the incidents of inadequate patient monitoring that result in patients filing insurance claims. As a result of fewer claims, the healthcare facility's insurance premium will be reduced, thereby saving money.

OPERATING A REVENUE CENTER

Every unit in a healthcare facility operates like a business—and the objective of every business is to have excess earnings at the end of the fiscal year. To achieve this result, the Nurse Manager must manage the unit as a business by making sure actual revenue and expenses follow the adopted budget for the unit.

The budget can be used as a guide to making fiscal decisions for the unit. Divide the budget in 12 month increments, based according to when income and expenses are expected. For most budget items, a Nurse Manager simply divides the budget amount by 12 to arrive at a monthly budget. However, the amount for a budgeted item can be distributed differently, if the Nurse Manager knows that there will be a seasonal shift in cases. A seasonal shift occurs when a demand for a service is higher in some months. For example, more orthopedic cases occur in the winter months in the Northeast because people slip and fall on snowy sidewalks. In this case, some orthopedic care expenses, such as traction devices or walkers, are shifted to the winter months.

A Nurse Manager should track actual amounts against budgeted amounts each month, note any variance between the actual and budgeted amounts, and explain why it occurred. A variance might be a sign of what is to come for the remaining months. Three or more consecutive variances in the same direction are considered a *trend*. A trend is used to forecast income and revenue for the remaining months. For the sake of the forecast, a Nurse Manager should assume that the trend of actual variances will continue at the same rate in future months. For example, suppose that as a Nurse Manager you budgeted for five colon cancer patients each week. However, for each of the first three months of the fiscal year, there were 15 patients—three times the expected demand. You could assume that this is a trend, and forecast that there will be 15 colon cancer patients for the fourth and fifth months also.

Once a budget is prepared and accepted, changes are not made to it. Everything is explained by looking at the variance. For example, with three times as many colon cancer patients, the unit's personnel expense line will be way over budget. This will be explained by noting the increase in the number of colon cancer patients, and the finance department will note that the hospital will expect an increase in revenues for this diagnosis that will cover the greater personnel costs.

Money is rarely moved from one category of the budget to cover another category. At the end of the year, the CEO presents the fiscal picture of the organization to the board of trustees by adding and subtracting every line item of the budget to arrive at a positive or negative bottom line. Within that bottom line, the senior management team knows, for example, that the high volume of surgical activities offset a loss in normal deliveries. However, the line items in the budget will show the details of how this situation came to pass: the labor and delivery budget will show an expense variance that is over the budgeted dollars, and the surgical unit will show a revenue variance that is also over what was budgeted.

By anticipating a change in demand, a Nurse Manager can adjust the budget to accommodate the expected increase or decrease. As a Nurse Manager, there are a number of ways to adjust a budget:

- Shift funds from one month to another to accommodate an unexpected seasonal shift. This is accomplished by moving funds within the same item. For example, if you budgeted $2,000 for linen supplies for April, and you expect a higher than normal demand, then you can take $500 from the remaining months' budget and add $4,000 to April's budget.

- Transfer funds from one account within the budget to another account, such as moving funds from the overtime budget to the linen supply budget. The transfer occurs within the same month.

- Ask senior management to transfer funds from another budget in the healthcare facility to the unit. This is a worse-case scenario that happens if actual expenses are higher than the overall budget. This usually occurs toward the last months of the fiscal year, when there are few places in the budget to transfer funds.

BREAKEVEN POINT

The *breakeven point* is when actual revenues are equal to actual expenses. Any revenue received beyond the breakeven point is considered excess earnings. The breakeven point reflects fixed and variable costs. *Fixed costs* are expenses that the unit incurs regardless of the number of patients cared for by the unit, such as costs for beds and facilities. *Variable costs* are expenses that vary depending on the number of patients in the unit, such as medication and I.V. tubing.

To manage the unit effectively, the Nurse Manager should allocate costs to each patient to ensure that fees charged to the patient cover expenses. It is straightforward to allocate variable costs to a patient, because these expenses are directly related to the patient's care, such as the cost of I.V. tubing used by the patient.

However, allocating fixed costs to a patient is not as obvious, because the relationship between a fixed cost and a patient's care isn't always apparent. For example, how much of the electrical cost should be allocated to each patient? Although you could go to extremes to calculate such an allocation, many healthcare facilities allocate fixed costs by dividing the fixed cost by the number of patients cared for by the unit. Let's say the unit's monthly electric bill is $2,000 and the unit cares for 200 patients a month. The prorated cost of electricity is $10 per patient. Somewhere in the fee charged to the patient is $10 to cover the cost of electricity. Calculating such an allocation would not be cost effective. Most healthcare organizations allocate expenses such as heat, light, and air conditioning on a per-square-foot basis. This cost will be bundled in a patient bill under some general heading, such as room and board.

A goal for every business is to operate with a low breakeven point, so that a large percent of revenue becomes excess earnings. One way to do this is to decrease the prorated fixed cost per patient by increasing the number of patients admitted to the unit each year. Suppose that the unit can care for 300 patients a month, but currently admits 200 patients. Fixed costs, such as the $2,000 electric bill, are the same regardless of the number of patients. However, the prorated cost per patient drops from $10, if 200 patients are admitted to the unit, to $6.67, if 300 patients are admitted to the unit. If $10 is included in the patient's fee, then the healthcare facility realizes excess income of $3.33 per patient. While this may seem a relatively small amount, consider the impact of all fixed cost allocations being reduced proportionally. This dramatically lowers the breakeven point for the unit.

Increasing the number of patients may or may not decrease variable cost—and lower the breakeven point—depending on the nature of the expense. Some variable costs will likely increase proportionally with the increase in patients. For example, more I.V. tubing is needed for 300 patients than for 200 patients. However, other variable costs will not rise proportionally. Let's say that an RN can care for seven patients. The cost of the RN is the same if there is one patient or seven patients.

Therefore, the cost of registered nursing increases in increments of seven patients, rather than for each patient.

BUDGET CONTROL

The unit's budget is a fiscal business plan for the unit. It tells the Nurse Manager what revenues the unit is expected to earn, and when those earnings will be realized. It also specifies the funds that can be spent to generate the expected earnings.

The Nurse Manager has the responsibility to operate the unit within the budget, and to do this the Nurse Manager needs to implement budget controls. A *budget control* is a procedure used to ensure that the unit achieves its fiscal goals.

Here are common budget controls that a Nurse Manager can use in a unit:

- **Inventory control** The unit should stock sufficient supplies to meet current patient care. For example, only a week's supply of I.V. tubing and other medical supplies should be available on the unit in an effort to avoid overstock, which results in overpurchasing of items.

- **Waste control** Identify situations that result in wasting supplies, and then devise ways to avoid those situations.

- **Purchase control** Make sure there are adequate funds in the budget before incurring the expense. Transfer funds before making the purchase, if necessary.

- **Maintain current budget records** Daily records should be kept of expenses. Reconcile the daily records with monthly statements received from the accounting department. This is like maintaining your check register and then comparing it to your monthly bank statement.

- **Monitor trends** Analyze the variance report each month to determine the percentage difference between the actual and budgeted amounts. This is your early warning sign that the unit is going into fiscal crisis.

Summary

A healthcare facility is financially organized into accounts grouped into revenue centers and cost centers. A revenue center is an account used to record income. A cost center is an account used to record expenses. Each department of the healthcare facility is assigned a cost center, and some departments are also assigned a revenue center, depending on the nature of the department. Revenue centers and cost centers contain actual revenue and expenses.

A budget is an accounting document that contains the healthcare facility's expected revenue and expenses. The healthcare facility's budget is divided into budgets for each department. Revenue-generating departments have both a revenue budget and an expense budget. The revenue budget specifies the amount of revenue that healthcare facility administrators expect the department to generate. The expense budget contains expected expenses for the department. Non-revenue generating departments only have an expense budget.

Revenue centers and cost centers are subdivided into a standard series of subaccounts referred to as the chart of accounts. Each subaccount is further divided into additional subaccounts. Each subaccount is assigned a unique number and description. The chart of accounts is used in budgets, revenue centers, and cost centers.

Budget planning is a collaborative process to devise the healthcare facility's budget based on review of the strategic plan, recent experience, and a determination of the impact of changes in the industry.

There are two categories of expenses estimated in budgets: operational expenses and capital expenses. An operational expense is incurred to operate the healthcare facility for the year. A capital expense is for a long-term asset.

The Nurse Manager must manage the unit as a business by making sure actual revenue and expenses follow the adopted budget for the unit. The Nurse Manager has the responsibility to operate the unit within the budget, and to do this the Nurse Manager needs to implement budget controls. A budget control is a procedure used to ensure that the unit achieves its fiscal goals.

Quiz

1. When actual revenues equal actual expenses, the unit is said to be:
   (a) Profitable
   (b) At its breakeven point
   (c) At capacity
   (d) None of the above

2. In what budget would you expect to find renovation expenses?
   (a) Capital budget
   (b) Operating budget
   (c) Overhead budget
   (d) None of the above

3. The healthcare facility's strategic plan:

   (a) Specifies tactics to negotiate union contracts

   (b) Details the organization's long-range plans

   (c) List ways to defend against litigation

   (d) None of the above

4. An increase in patient volume is expected to decrease the breakeven point.

   (a) True

   (b) False

5. Zero-based budgeting is a technique requiring the Nurse Manager to justify every item in the budget.

   (a) True

   (b) False

6. What determines whether the healthcare facility needs a short-term loan?

   (a) Level of patient care

   (b) Nurses' contract

   (c) Cash flow

   (d) None of the above

7. A product line is a group of services provided by different units in the healthcare facility to care for a patient who has a specific diagnosis.

   (a) True

   (b) False

8. A fiscal year is:

   (a) Any consecutive 12 months

   (b) Always a calendar year

   (c) The number of days financial institutions are open for business

   (d) None of the above

9. A special-purpose budget is a budget used for one-time activities.

   (a) True

   (b) False

10. A budget control is a procedure used to ensure that the unit achieves its fiscal goals.

   (a) True

   (b) False

CHAPTER 10

Unions, Management, and Employee Relations

Mention the word union, and images come to mind of picket lines and brawny men with broken noses wielding long wooden ax handles and baseball bats at truckers trying to deliver a shipment to a manufacturing plant.

Rarely do nurses come to mind. But times are changing. The nursing shortage, increasing demand for healthcare by aging baby boomers, and efforts by hospital administrators to bring skyrocketing healthcare costs back to earth are pressuring nurses to do the once unthinkable—unionize.

A union gives nurses strength in numbers to level the playing field with hospital administrators when negotiating work rules, compensation, and, most importantly, the quality of care patients receive in the hospital.

This chapter introduces you to collective unions, collective bargaining, and how to manage a staff of union and nonunion nurses.

The Changing Labor Environment

As the cost of healthcare escalates at an alarming rate, employers who underwrite healthcare expense for their employees are pressuring medical insurance companies to find ways to lower premiums. Federal and state governments who pay the medical care for the elderly, disabled, and poor are also seeking ways to lower the cost of medical care.

As a result, healthcare providers and healthcare facilities are forced to economize and lower their rates to an affordable level. In doing so, all levels and types of healthcare managers, including Nurse Managers, devise a budget that provides quality healthcare delivered at a cost covered by fees reimbursed by medical insurers and government agencies.

Controversy plagues the healthcare industry, as healthcare managers and healthcare providers define quality and affordable healthcare. A healthcare manager's definition sharply differs from healthcare providers such as nurses, because each views healthcare in a different light. Management may have little experience directly caring for patients—and that experience may not be current. On the other hand, nurses view healthcare from the frontlines, caring for patients with little or no direct experience dealing with cost of that care.

In the midst of this controversy, healthcare facilities are experiencing a new kind of patient, one who is acute, requiring intense care for a relatively short time. Decades ago, a patient was admitted for acute illness and remained hospitalized as they recovered. This meant that a unit had a mix of patients, some requiring close monitoring care and others who were well on the road to recovery.

Complicating the situation is a nursing shortage at a time when aging baby boomers are expected to overwhelm the healthcare system with a need some say outstrips the capability of the healthcare system to deliver at an affordable cost.

In this turmoil, healthcare managers must make tough decisions that at times appear as uncaring for patients, as well as nurses and other healthcare providers. Faced with an increased patient load and a shortage of nurses, a healthcare manager might require a unit nurse to care for seven to ten patients per shift, and to work mandatory overtime.

For an entire shift, a nurse is responsible for the medical care of up to ten patients. The nurse needs to assess each patient, verify that the patient is prescribed the correct medications, administer medications, perform special treatment procedures, coordinate with physicians, pharmacists, and other healthcare providers, and document the patient's condition and the care that is given to the patient. Along with this, the nurse provides emotional support to the patient and the patient's family. Failing to do any of this jeopardizes the patient's well-being and exposes the nurse and the healthcare facility to legal action.

Nurses have their own needs, such as restroom breaks and meal breaks—lower priorities that many times are not addressed. Furthermore, mandatory overtime intrudes on the nurse's off-duty life. While the healthcare manager leaves work at the same time each day, the nurse is never sure if, and when, they can go home.

It doesn't take long before a nurse feels abused by the healthcare system. The patient's physician spends a few minutes with the patient, writes medical orders, and then leaves nurses to carry out those orders and provide the patient with round-the-clock care. The healthcare manager gives nurses assignments and then has little direct involvement in the care of the patient. And, adding fuel to an already boiling situation, some physicians and healthcare managers treat nurses condescendingly and not as highly trained professionals who are at the forefront of delivery of medical care.

The turning point for many nurses comes when they look at their paycheck and compare it to healthcare administrators and others in the healthcare industry who don't literally hold the lives of up to ten patients in their hands each shift.

LABOR LAWS

Concerns about unreasonable work rules, adequate compensation, and an erosion of quality patient care have encouraged some nurses to organize into a collective bargaining unit to negotiate terms of employment with their employer.

A collective bargaining unit is commonly referred to as a labor union. In the manufacturing economy in the 1800s and the early 1900s, employers unilaterally established work rules that focused on producing goods at the lowest possible cost. Those rules imposed a harsh working environment that ultimately led to worker protests.

In 1935, Congress passed the National Labor Relations Act, which was also known as the Wagner Act. The National Labor Relations Act defines procedures for employees to freely choose a collective bargaining unit. Workers have the right to unionize and negotiate terms of employment with their employer.

The industrial union movement spread to the service area of the economy, including the healthcare industry. Healthcare workers could form a collective bargaining unit and have a say in rules that affected their jobs. However, management of not-for-profit hospitals felt that dealing with unionized workers placed them at a disadvantage and hampered their goal of servicing the community. Workers could impose demands that would divert funds from helping the sick, many of whom could not afford to pay for healthcare.

In 1947, Congress passed the Labor-Management Relations Act, commonly called the Taft-Hartley Act, which restricted some union activities and prohibited healthcare workers of not-for-profit hospitals from organizing into a collective bargaining unit. It took 27 years before Congress gave workers at not-for-profit hospitals the right to unionize, when in 1974 Congress amended the Taft-Hartley Act and removed the not-for-profit hospitals' exemption. All healthcare workers gained the right to form a collective bargaining unit.

Provoking Unionization

Collective bargaining units are formed when employees perceive that their employer maintains an unreasonable working environment. An unreasonable working environment includes a wide range of factors, such as work schedules, workplace safety, work rules, and compensation. Anything that is perceived as unfair by a group of employees can provoke them into forming a collective bargaining unit to change their working environment.

Once a collective bargaining unit is organized, the employer must negotiate all work rules with its representatives, which can be a time-consuming and tedious give-and-take process that is not conducive to responding quickly to changing healthcare needs of the community. Therefore, it is wise for the Nurse Manager whose workers are not unionized to resolve any contentious issues immediately before wide dissatisfaction develops among the staff. It is better to find a fair resolution to a concern, than to ignore it and have the issue fester among the staff.

A Nurse Manager should look for these warning signs of employee dissatisfaction with the working environment:

- **An increase in staff turnover on your unit and throughout the hospital**   Some employees would rather move on to a more favorable working environment than complain about their current one.

- **Complaints from physicians related to patient care and responsiveness of staff nurses to physicians' requests**   An underappreciated nurse is less likely to be attentive to every request from a physician.

- **Patient satisfaction scores that are lower each month**   Concerns about the staff's personal issues can overshadow care given to patients.

- **An increase in grievances to you or to the HR department by the staff**   A grievance is a formal complaint made by an employee regarding a work-related issue. An increase in grievances indicates an undercurrent of dissatisfaction that has reached a point at which employees actively try to change the work environment.

- **Low morale among staff**   The staff has the feeling that no matter what they do, nothing changes for the better.

- **Increase in sick time**   The staff simply would rather be someplace other than in the work environment.

- **Increase in complaints about hospital policies**   A hospital policy establishes work rules for your staff. Complaints will rise when policies that are enforced are perceived to be unreasonable.

- **Dissatisfaction with compensation**   This occurs whenever a worker feels undercompensated when compared to the perceived compensation in other healthcare facilities, with other healthcare workers, or with comparable workers in general.

When a Nurse Manager sees any of these signs, they can do the following:

- Intervene immediately.
- Approach the worker who raised the concern. Hold a staff meeting if the concerns are widely held among staff members.
- Be a good listener and let them explain the problem without any interruption.
- Ask what led them to their conclusion.
- See the issue from their point of view.
- Acknowledge their opinion without patronizing them.
- Ask them what they feel is a fair and reasonable resolution.
- Set expectations by letting them know what you are authorized to change.
- State that you will bring their concerns to your management—then do so.
- Provide the staff with honest progress reports to show that you are sincerely maintaining an open line of communication with them.
- Don't be afraid to deliver bad news, because truth builds trust.
- Clarify any misperceptions immediately.
- Be an advocate for your staff and speak up at appropriate management meetings and voice their concerns. The staff will appreciate the fact that you consider them a vital part of the healthcare team.

Unionizing

Employees who want to form a collective bargaining unit must follow procedures that are established in the National Labor Relations Act. The initial step is for 30 percent of the employees to sign a statement that they wish to explore collective bargaining for the healthcare organization, which they then file with the U.S. Department of Labor.

Once the statement is filed, organizers from established unions present employees with proposals to have their union represent them in negotiations with their employer. An election is then held, during which employees vote for the union of their choice. The union that receives 50 percent of the vote plus one additional vote is declared the winner and will represent the employees. Usually it is only one union vying to represent the employees, and if the percentage is not achieved the union is defeated. The same process is followed for employees to change or drop their existing union.

Campaigning occurs during the period before the election. Each union organizer tries to convince employees to vote for their union. Likewise, the employer campaigns against unionizing. Workers who support their employer vote no for any union. Campaigns are similar to political campaigns, with each candidate and the employer presenting facts supporting their cause. At times, campaigns can become downright dirty, where half-truths are told to mislead employees, or vicious rumors are spread to place the opposition in an unfavorable light.

Campaigns pit employee against employee, and employer against employee, and foster a sense of mistrust and pain among colleagues and friends. Throughout the campaign, the Nurse Manager must make sure that quality patient care is maintained. Any ill feelings must be defused immediately, and the staff must be refocused on the goal—caring for the patient.

Change is upsetting, but can also be beneficial. A collective bargaining agreement amicably addresses many issues that led to employee dissatisfaction and low morale. The Nurse Manager's responsibility is team building, by partnering with the staff and the collective bargaining unit to give patients the best possible care.

Pros and Cons of a Union

The election of a collective bargaining unit dramatically changes the work environment, because work rules are defined in a legal contract between the employer and the bargaining unit representing employees. Neither the employer nor an employee has the right to unilaterally change or ignore those rules.

On the positive side, the contract specifies overtime rules, floating to a different unit, reduction in shift rotation, flexibility in staffing, continuing education, tuition reimbursement, grievance and disciplinary procedures, and other potentially contentious issues.

On the negative side, some rules can be discouraging and impersonal. Promotions and transfers are likely based on seniority rather qualifications. A nurse can be discouraged by her peers from performing unassigned tasks, even if she volunteers to do so. The Nurse Manager has reduced discretion in interpreting agreed upon work rules. For example, the Nurse Manager can't give a nurse time off to care for personal matters, unless the action is in some way in keeping with the authorized collective bargaining agreement. And members of the bargaining unit have a new expense—union dues. Union dues are used to pay for negotiators, lawyers, and expenses to run the union.

NURSES UNIONS

Until 1974, the Taft-Hartley Act did not allow employees of non-profit organizations to be represented by a union. However, before 1974 nurses were permitted to be represented by an association, the first of which was the California Nurses Association. A nurses' association was not considered a union, although it negotiated work rules and compensation for its members.

Since the 1974 amendment to the Taft-Hartley Act, nurses unions have developed as part of state nurses' associations. Some nurses are also represented by the National Union of Hospital and Health Care Employees (NUHHCE) and the Service Employees International Union (SEIU). Historically, nurses haven't unionized because a breakdown in negotiations could lead to a strike. A strike isn't simply withholding work. It is patient abandonment, something nurses don't do. Seventeen percent of RNs are represented by a nurses' association or nonprofessional union. However, this is changing as a greater number of nurses seek representation.

Typically, professionals don't join a union, because the prestige of the profession is sufficient to secure acceptable working conditions. The increasing pressure to make the healthcare industry economically efficient is changing this view. Some nurses and other healthcare providers feel that they need the strength and protection of a union when dealing with their employer.

IS NURSING A PROFESSION?

Professionals don't unionize, but is nursing a profession? This has long been debated. A professional undergoes a long period of special training to develop valued expertise, which gives them autonomy to make independent decisions.

Many professionals enter into a client-provider relationship, such as a physician and his or her patient. Some professionals are self-employed, and those who are not are salaried employees. All adhere to a standard code of ethics of their profession.

In contrast, nurses don't work autonomously. They are hourly employees of a hospital or other healthcare provider, and take a more supportive role rather than primary role in patient care. That is, they provide care ordered by a physician. Training can be a minimum of two years for a nurse, compared with four to eight years in other professions. However, the majority of nurses consider themselves to be members of a profession.

Collective Bargaining

Collective bargaining is the process an employer and a collective bargaining unit use to agree on work rules and compensation. Once employees elect a collective bargaining unit, union negotiators meet with representatives of the employer to draw up a *collective bargaining agreement*, a legal contract that specifies work rules and compensation for members of the collective bargaining unit.

Each side creates a list of proposals, commonly referred to as demands. It is from this starting point that negotiating tactics are used to get the other side to agree to all or some of the proposals. If both sides can't reach an agreement, either side can declare an impasse and request the services of a *mediator*, a disinterested third party who tries to reach a compromise that can be agreed to by both sides. A mediator must use the power of persuasion because the mediator doesn't have any power to force an agreement.

Let's say it's the eleventh hour of negotiations and nurses are on the verge of a strike. Nurses want a 7 percent raise, and management is willing to give a 5 percent increase. The mediator might calculate the value of the across-the-board 2 percent difference to be $25,000. The mediator then privately asks management if it is worth $25,000 for the cost, disruption of service, negative publicity, animosity that lingers after a strike, and the aggravation that a strike will bring. Likewise, the mediator privately asks representatives of the bargaining unit if it is worth $25,000 to risk patient lives, cause animosity among the staff, receive negative publicity, and lower the image of nurses from a profession to a union worker in the eyes of the public. In essence, the mediator provides a powerful motivator to compromise.

There are other times in negotiations when there is an honest disagreement over facts of a situation. Rather than debate the issue, they'll hire a *fact-finder*, a disinterested third party whose sole purpose is to determine the facts of a situation. Suppose that the representative of the collective bargaining unit claims that the nurse-to-patient ratio is the highest among similar size hospitals in the region. Management disagrees. Both sides agree to bring in a fact-finder to research the nurse-to-patient ratio of other hospitals to determine the facts.

Another option, once an impasse is reached, is for both sides to agree to binding arbitration. *Binding arbitration* is a process where each side agrees to let a disinterested third party, called an *arbitrator*, determine terms of the collective bargaining agreement. Each side has the opportunity to present a proposal to the arbitrator. The arbitrator can accept either proposal, accept a mixture of both, or reject the proposals and come up with new terms.

Once an agreement is reached, and signed by both parties, the Nurse Manager and members of the collective bargaining unit are required to abide by the terms of the collective bargaining agreement.

SCOPE OF COLLECTIVE BARGAINING AGREEMENT

The collective bargaining agreement can have a wide scope covering issues that are of interest to management and nurses. These include mandatory versus voluntary overtime, input into decisions that affect patient care, opening lines of communication between nurses and management, sharing a balance of power between nurses and hospital administrators, and, of course, salary and benefits.

However, there are areas that are beyond the scope of a collective bargaining agreement. For example, terms of the agreement cannot be counter to laws and regulations of government agencies. Likewise, the agreement must abide by previously adjudicated issues. For example, suppose that as a result of litigation a court ruled that the hospital must have a nurse-to-patient ratio no greater than 1:5. The collective bargaining unit cannot agree to a ratio higher than 1:5.

Negotiating

In the months leading up to renewal of the collective bargaining agreement, nurses who are members of the union form a negotiating committee that is responsible for representing the union at the bargaining table. The negotiating committee listens to the concerns of union members and formulates a dream list of proposals regarding current work rules and an ideal new compensation package. They also whittle these down to the minimum that they'd accept.

The negotiating committee typically enlists the services of legal and financial experts who can guide them into crafting realistic proposals. They also usually hire a professional negotiator who helps develop a negotiation strategy and handles face-to-face negotiations with management.

Likewise, management also forms a negotiating committee, which often includes a Nurse Manager and other representatives of management. They too have a team of legal and financial advisors, and a professional negotiator, to help formulate their proposals.

Both committees meet and exchange their dream proposals. Sometimes this exchange occurs by mail rather than in-person. During the next several weeks, each negotiating committee reviews the other's dream proposals and places each proposal into one of three categories: acceptable, negotiable, or unacceptable.

Professional negotiators for each side take over and execute their strategy. Sincere negotiations usually occur hours before the expiration of the current collective bargaining agreement. It is at this point when negotiators reveal their whittled-down list of proposals and go head-to-head working out terms of an agreement.

Sometimes neither negotiating committee is directly involved in negotiations. Instead, the professional negotiators act as go-betweens presenting new terms to the negotiating committees. Once both sides agree to the terms, the negotiators, who might also be attorneys, write the terms of the agreement in a tentative agreement, which is signed by both negotiating committees. A tentative agreement contains terms that will be included in the final collective bargaining agreement. The tentative agreement is then presented to union members and the board of trustees for the hospital. Each side must vote to accept the agreement. They may also vote to reject the agreement, in which case it is back to the negotiating table.

Once the tentative agreement is accepted, lawyers for the union and management write the collective bargaining agreement, which is then signed by authorized representatives of the union and the hospital.

The Collective Bargaining Agreement

Many collective bargaining agreements contain the same key provisions, although the terms of each are unique to the agreement. These key provisions are as follows:

- **Wages**   Employees covered by the collective bargaining agreement are categorized by skill set and longevity. Wages are then assigned. For example, a starting RN with a BS degree will receive a specific wage, and may receive a different wage in the second year.

- **Seniority rights**   Seniority rights are used to reduce or eliminate opportunities to give an employee preferential treatment. Opportunities are granted to employees who have been with the hospital the longest. Seniority rights are typically used for layoffs, vacation accrual, shift rotation, weekend duty, transfers, and promotions.

- **Grievance resolution**   A grievance is a claim that either an employee or management violated the collective bargaining agreement. Grievance resolution is the manner in which the grievance is adjudicated.

- **Patient care**   The collective bargaining agreement may define the role a nurse has in making decisions about the care of a patient.

- **Union membership**   This provision determines which positions are covered under the collective bargaining agreement. Some agreements have a "forced union state" provision that requires all employees who hold a specified group of titles to be members of the union. Those employees must pay union dues, regardless of whether they want to join the union or not. Other agreements have an "open membership" provision that leaves it up to the nurse to decide whether or not to join the union. There are 21 states called "right to work" states that prohibit closed shops; www.nrtw.org/ rtws.htm gives a map of "right to work" states.

- **Assignments**   An agreement may have a provision that enables a nurse to object to an assignment, but the nurse must care for the patients once they accept the assignment. State boards regulate how this can be handled. For example, if reassigned to a unit that the nurse is unfamiliar with, the nurse can report to the unit and state the skills and nursing care that he or she feels they can safely provide. Management can take disciplinary actions if the nurse abandons the patient.

- **Staffing requirements**   This provision specifies the staffing requirements for each unit. However, this provision cannot overwrite staffing requirements specified by Medicare, the Joint Commission on Accreditation of Healthcare Organizations, and/or state regulators.

- **Overtime**   With a nursing shortage, and increased demand for healthcare workers, management may need nurses to work more than their normal shift. The overtime provision describes how overtime is allocated.

- **Pension or retirement packages**   This provision specifies retirement plans that are available to members of the collective bargaining unit.

- **Healthcare benefits**   This provision defines the health benefits for members of the collective bargaining unit.

- **Workplace hazards**   This provision specifies rules for ensuring that the workplace is a safe environment for nurses.

On Strike

A collective bargaining agreement is a contract between employees who belong to the union and the employer. As with any contract, there is a termination date, at which point the agreement is no longer valid.

Most collective bargaining agreements are renewed before the current agreement terminates. However, sometimes the termination date passes without the agreement being renewed. In those situations the union has two choices: extend the existing contract and continue bargaining in good faith, or walk off the job while negotiations continue.

At the point when contract talks reach an impasse, the union can vote to strike. The hospital must be given a ten-day notice before the strike takes effect to give management time to initiate its strike plan. This is referred to as a ten-day cooling-off period.

During this period, management reduces the hospital's census by discharging patients who can go home; transferring patients to other healthcare facilities; and reducing new admissions. Management staff is reassigned to provide care for the remaining patients. Arrangements are made to hire temporary nurses at a very high premium to staff units during the strike.

It is also at this time when a mediator is brought in to get talks back on track. All parties work very hard to prevent a strike.

Contract Signing

Once the collective bargaining agreement is ratified and signed, both management and union members need to reestablish a rapport so that operations can get back to normal. This process begins when senior management meets with union leaders to show support and goodwill to collectively address issues that arise in caring for patients.

Management needs to reassure union leaders that there will be no retaliation by management against union members, and that management will have only positive things to say about the union. The objective is to have everyone focus on the future and not the past.

The collective bargaining agreement is distributed to union members and management, and serves as the guideline for operating the facility. Any proposed resolution of contentious issues must be implement quickly to ensure that management intends to live up to all the terms of the collective bargaining agreement.

Senior management must also meet with other levels of management to explain contract changes and changes in work rules. Likewise, senior managers must meet with nonunion employees to explain work rule changes. The Nurse Manager must be familiar with the details of the agreement and use it as a rulebook.

The Workplace Changes

As a Nurse Manager, expect material changes to the way you manage your unit whenever a new collective bargaining agreement is signed. Avoid a win-lose attitude. Approach these changes positively, even if you are personally opposed to them. It is your responsibility to implement change. Remember that the change is new for you and your staff, and will be a learning curve before everyone is comfortable with new procedures. Make this known to the staff so they don't misread any miscues as management's lack of desire to implement the change.

Maintain a collegial work environment by focusing attention on patient care. Keep unionized employees content and productive. Treat the staff as equals. Union members are treated the same as nonunion members.

Remain open, inclusive, compassionate, and caring toward staff without favoritism. Apply rules uniformly and consistently. Remember, the union in not going away, so plan to get along with it.

Discipline, Suspension, and Terminations

The collective bargaining agreement details disciplinary procedures against an employee who violates terms of the agreement. It is critical for the Nurse Manager to have a thorough understanding of these procedures before contemplating action against a member of the union.

These procedures are designed to protect the employee by giving them a hearing in which they or a union representative can defend charges brought by management, and let an impartial third party decide if the charges are substantiated. If the charges are found to be meritorious, then the agreement dictates the type of discipline that can be imposed on the employee. This usually involves an official reprimand, suspension, or termination.

As a Nurse Manager, before making an allegation, make sure that you have documentation to support your claims. If you do, then you'll be in a good position to correct the problem with the employee. However, without supporting evidence, your claims will likely be dismissed. When in doubt, consult the HR department and union contract for guidance. You may be required to put the employee on a remediation plan to give them the chance to improve.

Summary

Some nurses organize into a collective bargaining unit to negotiate terms of employment with their employer. A collective bargaining unit is commonly referred to as a labor union. Collective bargaining units are formed when employees perceive that their employer maintains an unreasonable working environment.

The National Labor Relations Act defines procedures for employees to freely choose a collective bargaining unit. The Taft-Hartley Act restricted some union activities and prohibited healthcare workers of not-for-profit hospitals from organizing into a collective bargaining unit. Congress gave workers of not-for-profit hospitals the right to unionize when, in 1974, Congress amended the Taft-Hartley Act.

The Nurse Manager should resolve any contentious issue immediately, before wide dissatisfaction develops among the staff. It is better to find a fair resolution to a concern, than to ignore it and have the issue fester among the staff. Look for warning signs of employee dissatisfaction with the working environment.

As the initial step to forming a collective bargaining unit, 30 percent of the employees must sign and file with the U.S. Department of Labor a statement saying that they wish to explore collective bargaining for the healthcare organization. Organizers from established unions present proposals to have their union represent the employees in negotiations with their employer. The union that receives 50 percent of the vote plus one additional vote represents the employees. Nurses are represented by state nurses' associations, or by other unions such as NUHHCE or SEIU.

Collective bargaining is the process an employer and a collective bargaining unit use to agree on work rules and compensation. The result of collective bargaining is the collective bargaining agreement. A collective bargaining agreement is a contract between employees who belong to the union and the employer, and becomes the guide for operating units within the hospital.

Quiz

1. After a union votes to strike:

    (a) Nurses apply for positions in other healthcare facilities

    (b) There is an immediate walkout of nurses

    (c) There is a ten-day cooling-off period

    (d) None of the above

2. What is a grievance?

    (a) A claim that the collective bargaining agreement was violated

    (b) A course of action taken when an impasse is reached

    (c) A complaint nonunion nurses make to the union

    (d) None of the above

3. The process where each side agrees to let a disinterested third party determine terms of the collective bargaining agreement is called:

    (a) Mediation

    (b) Binding arbitration

    (c) Litigation

    (d) None of the above

4. A fact-finder is a disinterested third party whose job it is to determine facts of a contentious issue.

   (a) True

   (b) False

5. A mediator is a disinterested third party who tries to reach a compromise that can be agreed to by both sides.

   (a) True

   (b) False

6. A collective bargaining agreement sets terms for:

   (a) Overtime rules

   (b) Shift rotation

   (c) Tuition reimbursement

   (d) All of the above

7. A negotiating committee is responsible for representing the union at the bargaining table.

   (a) True

   (b) False

8. A tentative agreement:

   (a) Is a binding contract between the union and management

   (b) Must be approved by union members and the hospital's board of trustees

   (c) Is determined by a mediator

   (d) None of the above

9. Seniority rights are used to reduce or eliminate opportunities to give an employee preferential treatment.

   (a) True

   (b) False

10. The collective bargaining agreement details disciplinary procedures against an employee who violates terms of the agreement.

   (a) True

   (b) False

CHAPTER 11

Time Management

"There simply aren't enough hours in the day. No matter how hard I try I can't do everything I want to accomplish." Sound familiar? Well, join the club, because most of us say this daily. Yet there is always that one person who accomplishes everything that he or she set out to do—and then volunteers to help others.

How on earth does a nurse do it? The secret is time management. That industrious nurse doesn't complain or procrastinate, but instead uses self-discipline, time-management skills, and a lot of common sense to schedule activities in a way that would enable anyone to complete them on time.

The job of a Nurse Manager is to get things done correctly and on time with the tools and staff at their disposal. When you become a Nurse Manager, you don't have the luxury of letting things slip or be done halfway, because patients' lives depend on you completing your activities.

In this chapter, you'll learn proven techniques that successful Nurse Managers use to manage their time both at work and at home. This is one of the most important skills any nurse or Nurse Manager can have.

How Do You Use Your Time?

Here's a challenge. Starting tomorrow write a diary of your activities from the moment you wake up to the time your head hits the pillow. Your daily routine is naturally divided into activities. Some are short, like grabbing breakfast before running out the door, while others, such as getting dressed, take a bit longer. For each activity, briefly enter a description and the start and end time into the diary. Make sure that you list all phone calls, interruptions, and down time, if any, especially at work.

Take a few moments the following day to review your diary. You'll probably be surprised at how many activities you accomplished during the day. We tend to accomplish more than we remember. The diary describes your typical schedule, unless you performed an unusual activity during the day. The order in which you performed these activities is the implied priority that you assigned to those activities. For example, when you answered the telephone at work, you gave the call a higher priority than your current activity.

How many times during the day were there conflicts between two activities? Your diary will tell you, because you should have noted when one activity interrupted another, such as a call coming in while you were meeting with one of your nurses.

How did you handle these conflicts? Did you stop your discussion to answer the phone, or did you continue your discussion and let voice mail pick up the call? If you stopped your discussion and answered the phone, then you switched priorities. If not, then you maintained the original priority.

Like many of us, you probably let situations that arise during the day control your schedule, rather than you controlling your schedule. The phone call that interrupts your meeting is a good example. We are conditioned to stop whatever we're doing and answer the phone without considering the impact the call has on our current activity. In a sense, the phone caller barged into your schedule, pushing other activities aside without your approval. As a result, your current activity gets short-changed.

Step back for a moment and create an ideal schedule for yourself based on information in your diary. Place activities in order of priority. Follow your rearranged schedule and see if you accomplish more activities with less stress. It may take some explaining to the staff how you plan on handling your time. An open-door policy is great, but a tremendous time stealer. You can also have an open door at certain hours of the day, and appointments otherwise. You must be assertive in telling staff that you would like to visit with them when time can be arranged. Save the unbridled interruptions for true emergencies. One rule is: do not answer the phone if you're with someone who had an appointment. They gave you the courtesy of making an appointment—give them the respect of your undivided time.

It's a Matter of Time

There is truth in the old adage that says, ask a busy person to do something and it will get done. A busy person has the same 24 hours as everyone else, but knows how to balance his/her time between work and home, and between important things and those that are less important. In doing so, he/she reduces stress—for himself/ herself, her family, and her staff.

So how does wonder woman do all this? She manages her time by using these techniques:

- **Don't complain or procrastinate**   Take action instead, and you'll be well on your way to completing the task.

- **Divide large tasks into smaller ones**   It is easy to stay motivated when working toward goals that are achievable in a short period of time.

- **Analyze a task before performing it**   If the task doesn't fit your schedule, you may decide to let someone else do it so that the task is completed on time.

- **No interruptions**   Allocate time to perform a task and then perform it without being interrupted. Remember that you control your schedule.

- **Delegate and verify**   Delegate responsibility for completing a task if you can't do it yourself, and then verify that the task is properly performed on time.

- **Close your door**   Use your door to control your schedule. Open it to signal to your staff that they can drop by without an appointment to freely discuss any issue. Close it when you are working on a task and don't want to be disturbed. You may even put a sign on the door that states that you are working on a budget, etc., and leave a sign-up sheet for appointments on the door.

- **Say no**   Manage your schedule by turning down tasks that you know you don't have time to accomplish. Overbooking your schedule causes you to rush scheduled tasks, leading to poor-quality results.

Schedules

As a Nurse Manager, create a daily, weekly, and monthly schedule for yourself and for your staff. Include both work-related and home-related activities on your schedule to strike a balance between family and work. This is a great way to reduce stress, because you make time for yourself and your family.

Think of a schedule as money that you invest. The goal is to achieve the highest return on your investment. In the case of a schedule, you want to invest your time and your staff's time to return the most productive result. In order to reach this objective, you must thoughtfully choose activities for your schedule.

Base your decision on the scope of the activity. The scope of the activity consists of factors related to the activity, such as duration, start and end dates, complexity, staffing, and tools needed to perform the activity.

Perform a reality check by asking these questions when evaluating an activity for your schedule:

- Are expectations realistic? Make sure everyone involved in the activity has realistic expectations of its outcome before taking on the activity.

- Do you have time to perform the activity? Your and your staff's plates might be full, preventing you from undertaking the activity. Therefore, you cannot place it on your schedule.

- Do you have the skill sets and tools to perform the activity? If the scope is beyond your capabilities, then don't place the activity on the schedule.

If you cannot schedule an activity for one of these reasons, notify your supervisor and present him or her with the scope of the activity and your rationale for refusing to accept it.

Sometimes you can quickly assess an activity and determine where it belongs on your schedule. Other times you won't have all the information to make this assessment. In these situations, determining the scope of an activity becomes an activity itself and is added to your schedule.

PRIORITIZING

You must find a place in the schedule for activities that you accept. To do so, take these steps:

1. Break down the activity into smaller tasks.

2. Identify the duration for each small task and the skill sets needed to perform each task.

3. Determine times that you will personally be involved in each task, such as assigning the task to a staff member and supervising their performance.

4. Place each small task on your schedule and your staff's schedule.

5. Prioritize each small task.

Prioritization is the act of determining the order in which scheduled activities are performed—similar to performing triage in an emergency. To a large extent, prioritization is applying good common sense. However, there are a few pitfalls to avoid:

- Don't prioritize based on who yells the loudest, or who appears to be powerful inside or outside your organization. Let the facts of all the activities decide the priority instead.

- Don't assume that all tasks that make up an activity have equal priority. Some tasks have a higher priority than others within the same activity. For example, assessing the patient's condition and collecting insurance information are both tasks performed when admitting a patient to the emergency room. However, assessing the patient has the higher priority.

- Don't assign the same priority to two or more activities. Each activity, or task of an activity, has its own priority. If two or more activities seem to have the same priority, then assign them to different staff members.

- Make sure urgent activities, such as investigation of a medication error, receive the highest priority.

- Review end dates for an activity or task of an activity to help determine its priority.

- Be sure to schedule blocks of time for recurring activities, such as reading e-mails and listening to telephone messages.

LEVEL OF DIFFICULTY

Although we tend to generalize activities, each activity is unique and has its own level of complexity that makes performing the activity challenging for some and easy for others. As a Nurse Manager, it is very important to identify the complexity of an activity as it relates to the person who performs the activity.

Suppose you were told to manage the emergency room. Could you meet this challenge? It depends on your emergency department nursing skills and your managerial background. This could seem an insurmountable task if you had little or no experience in the emergency room. Yet a Nurse Manager with previous ER and managerial experience can quickly take control and successfully manage the emergency department.

If you lack the experience to perform an activity, then ask an experienced colleague for advice on the complexity of the activity and the skill set required to perform it. Don't assume that you have to perform the activity yourself.

You can always delegate the activity to a qualified member of your staff; however, it is your responsibility to identify the complexity of the activity and the skill set necessary to perform it.

SETTING DEADLINES

There might be one or more deadlines that span days, weeks, or months depending on the activity. There is one overall deadline that occurs when the activity successfully terminates and there may be deadlines for each subactivity.

Consider deadlines when fitting an activity into your schedule. Try to allocate a continuous block of time for each subactivity, rather than schedule interruptions by other activities. In this way, you can complete a subactivity within the deadline without interruption. Be sure that deadlines are realistic. All too often deadlines are set to motivate the staff, rather than to meet the needs of a situation.

Establish milestones and deliverables as a way to show progress if the deadline is greater than a month. A *milestone* is an easily recognizable accomplishment. Let's say that an activity is to restock all ten rooms of the unit. A milestone is reached when the fifth room is restocked. A *deliverable* is something related to the activity that is turned over to the person who requested it. Suppose five patients were discharged and the cleaning staff needs to refurbish those five rooms. When the first room is refurbished, it is "delivered" to the staff for use by another patient.

GETTING ORGANIZED

List on your daily schedule the information you need to complete each activity, and then prepare the information a day ahead of time. Use a sticky note to clearly identify each piece of information and attach it to the document itself. Make sure to specify the type of information it is, the activity it is for, and how you intend to use it. For example, you might label a document as "last week's unit census for weekly review to assess staff performance."

You should organize information needed to complete your daily schedule. There are various schemes that you can use to do this, such as using color-coded folders, with a different color for each activity. Place information for each project in the appropriate folder so that you have all the information at your fingertips when performing the activity. A calendar of some sort with daily time slots is vital. Some Nurse Managers prefer a pocket personal computer device, because it contains a calendar, contacts, important documents, telephone numbers, etc. Others prefer a Day-Timer type of notebook. Determine whatever works best for you. Don't be afraid to try different ways of keeping track of your days until you find the one that best fits your style.

Time Wasters

Be on the lookout for the following nonproductive activities and try to eliminate them where possible:

- **Unnecessary meetings**   Regularly scheduled meetings are without a doubt a popular time waster. Initially these meetings are planned to update everyone on the latest information concerning their area. However, meetings are continually scheduled when there is nothing to update. This leads to participants spending time prior to the meeting coming up with things to say at the meeting. Call a meeting only when you have something to say or to discuss. Use e-mail or use other means of communication as appropriate, to deliver your message.

- **Drop-in visits**   An unscheduled visit interrupts a scheduled activity and, many times, requires you to schedule another visit because you are unprepared to discuss the topic.

- **Acting with incomplete information**   Too often action is taken based on incomplete information. As a result, you have to take a different action once all the information is at hand. The initial effort is wasted. Always step back and ask yourself if you have all the information you need to take action.

- **Unclear objectives**   The goal must be clear to everyone involved in the activity. If it isn't, then time is wasted trying to determine the clear objective of the activity.

- **Unclear communication**   A great plan is useless if it cannot be clearly communicated to those who have to enact it. Make sure to use the most appropriate means to communicate your plan, and then verify that your message was accurately received.

Productive Meetings

Meetings can be productive if you follow a few basic guidelines:

- Create a limited agenda for the meeting that focuses on no more than three related topics that can be thoroughly discussed within an hour. Clearly state on the agenda the expected outcome of the meeting, the start and end times, and the list of participants.

- Distribute the agenda a week before the meeting. This gives everyone time to prepare to discuss the issues and, in some cases, suggest alternative participants who could better address the topics.

- Schedule a meeting for midmorning rather than for the start of the workday. This gives everyone time to settle in and take care of a few chores before attending the meeting.

- Schedule a meeting for midafternoon rather than immediately following lunch. This gives everyone time to address any business issues and personal needs after lunch.

- Invite only participants who can speak about the topics of the meeting and give any necessary approvals. Avoid inviting a manager, if the manager's staff member is the one who can speak in detail about the topic. Likewise, don't invite the staff member, if only the manager can authorize a resolution regarding the topic.

- Keep the meeting to an hour in length. Participants tend to lose interest after the first hour.

- Stay focused on the agenda items. Insist that discussion of other topics be placed on the agenda for another meeting.

- Go to a meeting with an objective and leave the meeting having accomplished the objective. Don't reconvene the meeting.

- Start and end the meeting on time.

- Leave a 30-minute period between meetings. Avoid back-to-back meetings.

- Take minutes at the meeting and distribute them to participants. The minutes should clearly state who attended the meeting, any concerns that were raised, and who is responsible for actions related to the outcome of the meeting.

A LOOK INSIDE YOURSELF

A Nurse Manager must be decisive in making decisions; however, some Nurse Managers are indecisive and procrastinate, fearing that they'll make the wrong decision. It is critical to overcome fears, because otherwise no decisions will be made.

Develop the self-discipline to approach every decision with a positive, "yes I can" attitude. Follow a procedure that brings you to a rational decision based on the circumstances of the situation. You might make the wrong decision, but others following your thought process will do so too.

The following are some ways to become an effective decision maker:

- Avoid knee-jerk decisions.

- Understand the problem before making a decision.

- Separate fact from emotion and misperceptions.

- Determine when a decision must be made. Few decisions have to be made on the spot, which means that you'll probably have time to digest the facts, discuss the issue with a colleague or supervisor, and identify alternative solutions before making the decision.

- Identify factors that you'd expect to consider when making the decision, and then seek out those factors before deciding. Let's say that a staff member asks for a day off. Before granting it, you normally determine staffing for that day and whether the staff member has used their allotted days off. Therefore, you'd hold off on making a decision until you review those factors.

- Make a list of viable solutions and identify the advantages and disadvantages of each. Consider the impact each of these might have on other, future decisions. Your decision might be setting a precedent that will dictate how similar decisions are reached.

- Ask for help. You are not an expert in all factors that impact your decision. Therefore, you'll need to call in those experts for advice. Make sure they explain to you their basis for the advice that they give to you. Don't be insecure—surround yourself with people smarter than you, and they really will help you. If you are the only one that speaks, you will only know what you know. If you let others speak, you know what you know *and* what they know.

- Review your decision with your supervisor or colleague, if possible, to get a second opinion. Many times another person can see faults with your decision that are not apparent to you.

- Address both the factual and the perceived problem. For example, a patient might complain that the staff is giving another patient preferential treatment because the other patient is a physician. The physician might be receiving equal treatment; however, you should acknowledge how the patient arrived at this perception, and then assure the patient that they are receiving the same level of care.

- Deliver your decision in the most appropriate way. Some decisions can be simply conveyed verbally. Other decisions must be carefully worded and presented in writing, such as those involving a bargaining unit member. Your HR department can advise you on the proper means of communicating your decision.

- Anticipate repercussions. Even correct decisions can result in negative repercussions. Suppose you refuse granting a nurse a day off because of scheduling problems. Depending on the nurse's personality, you might expect the nurse to bad mouth you to the rest of the staff in an effort to portray you as a villain. You can anticipate this move and counteract it by preparing a memo to the staff thanking them for going above and beyond during the period of a staff shortage and a higher than expected census.

- Make note of your decision. Include your perception of the facts, the results of any investigation that you might have made into the situation, a list of all possible solutions, the advantages and disadvantages of each, and the rationale for selecting one of them.

- Refer to your notes when anyone discusses your decision. Sometimes months might pass and HR or your supervisor might ask you about it. Refresh your memory using the notes, rather than responding by memory.

Summary

To manage your time effectively, create a diary of your activities and use it to reconstruct your schedule. Then, look for areas where you can better manage your time. Don't complain or procrastinate; instead, divide large tasks into smaller ones and focus on completing those tasks.

As a Nurse Manager, avoid interruptions by using the door to your office as a signal to the staff that you are on an assignment and shouldn't be disturbed. Say "no" when you are asked to take on an activity that doesn't fit your schedule, or say "yes" and delegate it to your staff.

Create a daily, weekly, and monthly schedule for yourself and for your staff. Before placing an activity on your schedule, perform a reality check and evaluate the scope of the activity. Set expectations before agreeing to perform the activity.

Prioritize your schedule by focusing on the facts of the situation. Avoid assigning two activities the same priority, because they are performed one at a time. Highest priority is given to urgent activities.

Assess an activity's level of difficulty and the deadline, and then assign it to the staff member whose experience can ensure that the activity is completed on time. Consider establishing milestones and deliverables as a way to show progress.

The day before you begin, gather information needed to complete the activity and organize it so that you have it available at your fingertips and don't have to waste time finding it. Be alert to time wasters and avoid them at all cost.

Make sure that your meetings are productive and focus on addressing no more than three topics. Schedule a meeting only when you have something to discuss. Avoid regularly scheduled meetings.

Quiz

1. Set priority based on:

    (a) The facts of the activity

    (b) The position of the person who requested the activity

    (c) Your preference

    (d) None of the above

2. What is a milestone?

    (a) An easily recognizable accomplishment

    (b) A part of the activity turned over to the person who requested the activity

    (c) A completed activity

    (d) None of the above

3. What is a deliverable?

    (a) An easily recognizable accomplishment

    (b) A part of the activity turned over to the person who requested the activity

    (c) A completed activity

    (d) None of the above

4. Acting with incomplete information is a time waster.

    (a) True

    (b) False

5. The scope of an activity provides facts used to decide if the activity should be scheduled.

    (a) True

    (b) False

6. Placing both home and work activities on your schedule:

   (a) Reduces stress

   (b) Provides a balanced schedule

   (c) Ensures that you address your family's needs

   (d) All of the above

7. Distributing the agenda a week before the meeting can make for a productive meeting.

   (a) True

   (b) False

8. The best way to make a decision is to:

   (a) Understand the problem before making a decision

   (b) Separate fact from emotion and misperceptions

   (c) Determine when a decision must be made

   (d) All of the above

9. You should leave a 30-minute period between meetings.

   (a) True

   (b) False

10. An unclear objective is a time waster.

    (a) True

    (b) False

CHAPTER 12

Nursing Informatics and Measurement

Information overload … information overload. Imagine these words flashing inside your head as you try to sift through seemingly endless information to find the right piece to help you get your patient back on their feet.

There must be a better way—and there is, by using informatics and measurement to filter out unimportant data, leaving only the information you and other healthcare providers need to care for patients.

You'll learn about informatics and measurement in this chapter and see how, by combining them with computer technology, you can dramatically improve how you collect and use healthcare information.

What Is Informatics?

Informatics is the study of information and how information is communicated. Today we tend to associate informatics with electronic information managed by a computer application; however, informatics deals with both nonelectronic and electronic information.

For example, the Medication Administration Record (MAR) shown in Figure 12-1 is an application of informatics. Hospitals administrators had a problem: how to schedule patient medication and record when the medication was administered to the patient. The solution had to convey medication orders in a clear, concise way so that any nurse could safely use it to treat patients. The solution also had to clearly identify the nurse who administered the medication.

Concepts of informatics were applied, and the result is the MAR, which lists medications in the first column and the time medications were administered in the other columns. Each cell is of a sufficient size to write the initials of the nurse who administered the medication. At the bottom of the table is a box where the nurse writes their initials and signature to clarify whose initials appeared in a cell.

Informatics is used to record both administrative and healthcare activities throughout a healthcare organization, such as for employment applications, payroll, benefits, inventory tracking, and direct patient care, such as medical charts, lab reports, and MARs.

What Is Measurement?

Measurement is a process used to analyze empirical data collected by informatics. *Empirical data* is raw information. Think of informatics as a way to collect and present empirical data. Measurement is a way to make sense out of that information.

Let's say a physician wants to know the frequency that a patient received Demerol. The nurse reviews entries (empirical data) in the patient's MAR and concludes (analyzes) that the patient received a dose every four hours beginning at 10 A.M.

The Nurse Manager uses the combination of informatics and measurement to manage the unit. By measuring data collected using informatics, the Nurse Manager is able to understand what has happened, what is happening, and, by making an educated guess, what is likely to happen in the future. This is why many believe that a bachelor's, and preferably a master's degree, should be required for nursing management.

MEDICATION ADMINISTRATION RECORD

Consumer Name:

Attending Physician:

Month:

Year:

MEDICATION	HOUR	1	2	3	4	5	6	7	8	9	10	11	12	13	14	15	16	17	18	19	20	21	22	23	24	25	26	27	28	29	30	31

R = REFUSED   D = DISCONTINUED   H = HOME   D = DAY PROGRAM   C = CHANGED

REMEMBER TO RECORD AT TIME OF ADMINISTRATION

Figure 12-1   A Medication Administration Record is an application of informatics used to facilitate administrating medication to patients.

Electronic Informatics and Measurement

The present trend is to use computers to gather and analyze empirical data and to report the result to healthcare providers, who use it to quickly respond to changes in the patient's condition. This is happening in critical care units and elsewhere in the hospital where patients are connected to electronic devices that monitor their pulse, pulse oxygenation, heart rhythm, respiration, and blood pressure.

The nurse, or a technician, attaches transponders to the patient. A *transponder* is a device that transfers mechanical or chemical events into electronic pulses that are transmitted to an electronic device—a computer. These devices display empirical data as a number (pulse rate) or time-series data in the form of a graph (cardiac rhythm). Typically these devices have a few analysis tools. For example, a device may beep whenever a patient's vital sign exceeds a safe range, or when something is amiss with the intravenous pump.

There is also a trend toward increasing the capability of electronic devices used in healthcare to perform other analyses. Electronic lab equipment analyzes a specimen such as blood, and then reports the results as low, normal, or high. Some devices also list possible diagnoses based on probability.

There are also electronic devices that gather empirical data (informatics), analyze data (measurement), make a diagnosis, and treat the patient with little or no human intervention. The implantable pacemaker is an example of this. If the patient's heart begins to beat too fast or too slowly, the pacemaker will restore the heart to a normal rhythm. The automated external defibrillator (AED) has long been used to deliver a shock to a patient's heart and restore the beat. This requires human intervention. Newer technology, not requiring human intervention, has provided patients with an implantable cardioverter defibrillator that automatically delivers a shock to the patient's heart when it recognizes fibrillation. This is another example of one of those electronic devices.

Nursing Informatics

An endless amount of information must be gathered before a patient is treated and returned to normal daily activities. It is the responsibility of hospital management to devise a system for collecting accurate patient information and disseminating it in the clearest, most useful form to the healthcare providers who must care for the patient. This is an enormous undertaking.

This section describes several steps that a Nurse Manager can take to determine the information to collect and how to collect it.

DECIDE WHAT INFORMATION IS IMPORTANT

Design your informatics around types of patient encounters. Gather sufficient information to document the assessment of the patient, diagnosis, and treatment. For example, a patient presents in the ER with a broken arm. Is it important to know that the patient had two normal deliveries? Probably not, because this isn't relevant to the patient's current complaint. Yet that history is critical if the patient is in her third trimester presenting with abdominal pain.

DECIDE HOW TO GATHER THE INFORMATION

Design an efficient and convenient way to collect data and reduce data input errors. If forms are used, then design a form that flows in the way the healthcare provider normally acquires the data for each type of patient encounter.

Don't assume that computers are the best way to gather data, because in some situations computers are inconvenient. Computers are convenient to use at the admissions desk, where the admissions clerk can enter the patient's responses directly into the system. They are also ideal for the nurses' station, the pharmacy, and elsewhere throughout the hospital. Computers are inconvenient to use at the bedside. Laptop computers are too bulky, and handheld computers and cell phones are too small to be used for data entry. However, it won't be long before the world of information technology creates very useful handheld devices. A combination of paper and electronic data collection methods is the best data-gathering solution at this moment in time.

Anticipate that some healthcare providers are not comfortable using a computer and provide alternative means for entering data into the system.

Errors are dramatically reduced by carefully designing a form. Here's what you need to do:

- List questions to ask the patient in the order they will be asked. This ensures that the healthcare provider won't forget to ask required questions.
- Display the range of common responses in check boxes or a word list on the form. This saves time writing the patient's response, and ensures that the response is legible and uniform.
- Place questions in related groups on the form. This helps the healthcare provider focus on one topic of questions before moving on to another.
- Be sure to include follow-up questions that the healthcare provider should ask based on the patient's response to a previous question.

ASSESS CONFIDENTIALITY

Assess the confidentiality needs of information. Patient information is available on a need-to-know basis. Therefore, organize data so that only information the health-care provider or support staff are entitled to see is on the form.

Confidentiality considerations extend beyond data collection. Assess confidentiality requirements for the transcription of information from a paper form to the computer, for the storage of the paper form, and for who has rights to access the patient's computer information.

DECIDE HOW THE INFORMATION WILL BE ANALYZED

Find out what information hospital administrators, healthcare providers, and support staff need to do their jobs, and then decide how to collect that information. Many times, summary data rather than empirical data is required, such as the average length of a patient stay in the unit. Determine the empirical data and the formula applied to the empirical data to generate the summary data.

DECIDE HOW THE INFORMATION WILL BE STORED

Determine how to store the data. *Data storage* is the placement of information in a secure place where it can be retrieved within a reasonable time period by those who are authorized to access it. Paper forms are stored on the premises for a short period after being transcribed into the computer. Eventually, the paper forms are sent to a secure third-party facility, and kept there until they are no longer required for financial or legal purposes, at which time they are destroyed.

Electronic data is stored in the hospital's data center and linked electronically to computers throughout the hospital. Electronic copies of the data are stored off the premises at a third-party facility, where they too are kept until financial and legal requirements are satisfied.

Electronic Medical Record

Healthcare providers, government agencies, and medical insurers are leading the effort to create an Electronic Medical Record (EMR) that standardizes the informatics used for patient information across all areas of the healthcare industry.

An EMR is being rolled out in three phases:

1. Automate the paper chart. All charting is computerized, eliminating the need for the paper forms that are currently used to document care given to a patient.

2. Passive decision support. Provides an easy and fast way to retrieve information that is important to the current patient encounter. No longer does the healthcare provider have to leave the nurses' station or patient's bedside to locate patient information needed for the immediate treatment of the patient.

3. Personal EMR. An EMR will contain a tailored care plan based on the patient's diagnosis.

The Centers for Medicare and Medicaid Services (CMS) is spearheading an effort to create a national standard for an EMR. With funding from the Department of Health and Human Services (DHHS), CMS formed the Healthcare Information Technology Standards Panel, which is working with government agencies, academia, medical insurers, and healthcare providers to devise this national standard.

BARRIERS TO THE ELECTRONIC MEDICAL RECORD

A number of practical hurdles facing the healthcare industry hinder implementing an EMR. Financing is a major stumbling block. An EMR requires a substantial increase to the capital and operating budgets of hospitals at a time when there is an industry-wide effort to substantially reduce the cost of patient care. Each hospital needs to purchase and install computers, acquire the EMR computer program, and install networking cables to link the computers. There are also additional ongoing costs for electricity, software maintenance, and computer and software technicians.

Physician resistance is another barrier. Some physicians are not comfortable with interacting with a computer, and question the accuracy of computer-generated information. Hospital administrators must incur added expense and spend time to address these concerns before implementing an EMR.

Confidentiality of patient information is another roadblock to quick implementation of an EMR. Each hospital is held liable for any breach of patient confidentiality. All hospitals have a system in place that protects this information. Hospital administrators are leery of giving up a proven secure system for a new EMR, especially in the era of computer hackers.

It is for these reasons that only 10 percent of hospitals in the United States have started to implement a comprehensive EMR system.

Technology 101

Understanding technology is critical to designing and implementing systems that use informatics and measurements in healthcare delivery. This section provides an overview of how technology works.

Computers are at the heart of informatics technology. A computer is an electronic device that collects, stores, manipulates, and displays information. Computers are categorized as dedicated or nondedicated. A *dedicated computer* is designed to do one thing, such as a patient monitor that collects information about the patient's vital signs and then stores that data in memory, where a program manipulates it before displaying the information on the screen as a number or a graph. Because of the almost limitless capacity and capability of computers, patient information must be protected, even if it is not in the form of a traditional patient record.

Nondedicated computers are able to do many things by running a variety of applications. An *application* refers to nearly all the programs that you interact with on your computer.

A computer consists of software and hardware. *Software* refers to the set of electronic instructions, called a *program*, that tells the computer what to do, how to do it, and when to do it. *Hardware* is the term given to the monitor, keyboard, mouse, and all the circuitry and devices inside the computer case.

Computer hardware takes on various forms that include dedicated computers, such as ICU monitors, and nondedicated computers, such as laptops, desktops, and personal digital assistants (PDAs). All have the same basic internal components.

SOFTWARE

A computer can display something on the screen, read data from the keyboard, make decisions, and perform math as well as a lot of other basic actions. However, the computer needs a set of instructions, called *software*, to tell it when to do these things. Software is also referred to as a *computer program*. The computer programs that you use are called *applications* because they apply computer technology to solving a problem.

Application programs fall into one of two categories. A *custom application* is one written specifically for you. A *commercial application* is written to fill a common need and is available for purchase from a vendor.

Commercial application programs fall into one of two groups: *packaged programs*, such as Microsoft Word, that you buy over the Internet or at your local computer store, and *nonpackaged programs*, such as the hospital's EMR system, sold by the vendors who write them.

HARDWARE

Computer hardware is the physical computer and consists of input devices, storage devices, processing devices, and output devices. Although these terms might seem unfamiliar, you are well aware of these devices:

- **Input device**   Hardware used to collect information. Input devices include not only the keyboard, mouse, and microphone, but also the cables that carry electronic signals from the patient into the computer and cables that connect to other computers.

- **Storage device**   A place where information is held. Storage devices include disk drive, DVD drive, CD drive, and memory chips inside the computer, as well as USB flash memory that fits on a keychain or lanyard and plugs into the computer.

- **Processing device**   A computer chip used for manipulating data. This is better known as the computer's *central processing unit (CPU)*.

- **Output device**   Hardware that takes information away from the computer. Output devices include the monitor, audio speakers, printers, and cables that connect to other computers. Output devices also include the cables leading from the computer to the patient, as in the case of an AED, where the computer causes a charge to defibrillate a patient.

COMPUTER NETWORKS

Information is shared among computers using a *computer network*, which consists of a shared computer, called a *server*, a router, cables, and other computers, called *clients*, used to access the network.

Think of a computer network as a town. Client computers such as a desktop computer are the homes. The server is the post office. The router is the traffic cop standing in the center of town. Cables are the streets that lead from each computer to the router. Each computer and printer connected to the network has an address, called an *Internet Protocol (IP) address*, much like each house has an address.

A computer sends a request over the network when it wants to contact another computer. The request is like a letter that you send through the post office—but this isn't e-mail. Similar to a letter, the request has a recipient (to) address and a sender (from) address, except these are IP addresses instead of house addresses. The request travels along the cable (street) to the router (traffic cop). The router looks at the recipient's IP address and forwards the request along the cable leading to the recipient's computer or a printer (house).

A computer network within a healthcare facility is referred to as a *local area network (LAN)* because it is local to the healthcare facility. Some healthcare facilities also connect to a larger network that consists of other healthcare facilities. This is referred to as a *wide area network (WAN)*.

A LAN in a large healthcare facility is typically divided into smaller networks, called *segments*. Think of each segment as a neighborhood within a town, having a server, router, and client computers. A cable connects routers for these segments, enabling any client computer to send a request to a computer on its network segment or a computer located on any network segment within the LAN.

When a router receives a request addressed to a recipient on another network segment, it forwards the request to the router for that network segment, which in turn forwards the request to the recipient.

NETWORKED SERVICES

Client computers run application programs such as an ERM or patient health insurance data.

Nurses, physicians, and other healthcare providers use an application program to request information that is stored on a computer elsewhere on the network.

As an example of how networked services work, suppose that a nurse wants to display a patient's lab reports. The nurse makes the request using an application program on a client computer. The request might be sent across the LAN to a database server, a computer on the LAN that stores and manages information. The request contains the database server's IP address, the nurse's client computer's IP address, and a message. The message might say, "Give me the labs for patient 12345," but in words the database server understands. The database server looks in an electronic filing cabinet for patient 12345's labs, makes a copy of them, and creates a response. The response contains the lab results, which are sent back to the client computer that made the request. The sender's IP address on the request becomes the recipient's IP address on the response. The application program on the client computer removes the lab data from the response and displays it on the screen.

A LAN can have many different kinds of networked services besides a database server. For example, it might also have networked printers, enabling client computers to share a printer. The LAN is likely to have shared access to the Internet and its own e-mail service.

Networked services that are shared by all segments of the LAN are grouped into their own LAN segment called a *backbone*. The backbone segment typically contains the database server, e-mail server, and a shared connection to the Internet.

DATA INTEGRITY

Healthcare professionals and hospital administrators are concerned about the accuracy of data that is presented on the computer monitor. Data generated by a computer typically undergoes a verification process before being saved to a database server. The verification process applies rules created by the team who developed the application used to input the data into the computer.

A data verification process is unique to each application, but many of these processes use similar techniques:

- **Present a selection of valid choices**   This eliminates the opportunity for the healthcare provider to make an inappropriate choice. For example, the healthcare provider can check the appropriate dose from a selection of common doses.

- **Restrict the type of data that can be entered**   For example, if the data that is entered should be a number, the user is prevented from entering letters.

- **Conduct reasonableness testing**   This is performed to ensure that data makes sense. For example, the end date for treatment must be a later date than the start date. Numeric values should be within an acceptable range. An application program should question a pulse of 300.

- **Verify data instantly**   This is performed by comparing data supplied by the healthcare provider to data known to be valid, such as a patient ID number. Suppose a healthcare provider inadvertently enters a patient ID number of a male patient into an application program to schedule a hysterectomy. The application looks up the patient ID and displays a warning message pointing out the conflict.

SECURITY

Security concerns are paramount when it comes to patient information. Although press reports imply that electronic data is highly susceptible to attacks by computer hackers, data is protected by strong security measures.

Each healthcare provider, hospital administrator, and support staff person is issued a unique user ID and required to choose a password that meets security standards. These standards typically require that the password be at least eight characters consisting of a mixture of upper- and lowercase and alphanumeric characters. Easily guessed passwords, such as the name of the hospital or the user's name are not permitted. Further protection is provided by requiring the password to be changed regularly and not allowing previously used passwords.

Each user ID is granted permission to use specific applications and access certain data depending on their needs. This is referred to as a *user profile*. For example, a pharmacist has access to parts of a patient's EMR that pertain to the medication, but is prohibited from displaying other patient information.

The user ID and password are used to log into the hospital's computer network. The user ID and password are validated against a login database. Passwords in the login database are usually encrypted, which prevents anyone, including the hospital's Information Technology (IT) department, from knowing a user's password. If the user forgets their password, then the IT department assigns a new password that must be immediately changed. You must protect your password to prevent unauthorized access to computerized services. If an unauthorized person goes into the system using your ID and password to retrieve sensitive information (a patient's diagnosis, perhaps), and uses that information to cause harm to the patient, you will be accused of causing the harm or distress to the patient. You will have no proof of your innocence and the outcome could be the loss of your job. So be very careful.

Users are typically given three chances to enter a correct password before the user ID is automatically suspended. The user then needs to contact the IT department, which verifies the user's identity and then resets the user ID. This technique limits the opportunity for a hacker to guess the user's password.

As another precaution, users are automatically logged off if they don't use the computer for a specified amount of time (usually 15 minutes or so), depending on the hospital's security policy. This prevents someone else from accessing the computer if the authorized user steps away.

Interactions with the computer are usually recorded in a log. A log is like an electronic notepad that contains the date, time, user ID, activity, and the ID of the client computer that is used to perform the activity. Besides containing login attempts, logs also record the use of application programs and data accessed by the user.

The log is used as an audit trail to reconstruct events that might be associated with a security violation. For example, suppose that an unauthorized person tried to use the client computer at the nurses' station to log into the hospital's computer network. The unauthorized person observed the unit secretary logging in and attempted to memorize her user ID and password. The secretary stepped away for 20 minutes, during which time she was automatically logged out. The unauthorized person tried logging back in. The user ID worked, but after three failed attempts at entering the password, which was incorrectly memorized, the user ID was suspended and the unauthorized person left the unit. The unit secretary contacted the IT department after being unable to log back in when she returned to her station. This raised suspicions, and the IT department used the log to reconstruct events that occurred in the 20 minutes that she was away from the unit. The log provided the information that the secretary failed to promptly log out, and that an unknown and unauthorized person attempted to log on with the secretary's ID and password. The secretary would be chastised for walking away from her computer without logging out and the unit could be monitored. Is this a common scenario and is an unauthorized person, who may be an employee, frequently attempting this deception.

Electronic data is replicated automatically and stored in different locations. If a catastrophe hits the hospital's data center, the IT department implements contingency plans to use an off-site data facility.

BUILDING VS. BUYING APPLICATION PROGRAMS

Commercial applications can be purchased from a vendor or they can be built in-house by the hospital's IT department. Hospital administrators determine which alternative to choose based on need and financial resources.

Hospitals have similar but differing needs for application programs. For example, all hospitals need an EMR system that complements the hospital's existing patient medical record system. However, the existing patient record system is different in each hospital. Hospitals collect and store the same data, but each does so in a different way.

Purchasing a vendor application program is less expensive than building the application in-house, because the cost of building the vendor application is shared among all hospitals that purchase the off-the-shelf application. However, for almost all such vendor application programs, hospitals are required to adopt the vendor's way of collecting and storing information. As a result, additional time and money is needed to convert existing information to the vendor's system.

Furthermore, only the vendor can modify their application. The hospital's IT department cannot perform needed modifications. In addition, the hospital is expected to pay the vendor to maintain the program. The hospital becomes totally dependent on the vendor for support of its application. If the vendor is unable to support the program, the hospital needs to find an alternative application, unless special arrangements are made with the vendor.

Building an in-house application program is expensive, because the hospital underwrites the entire expense. However, the custom application complements the existing way the hospital collects and stores information and the hospital has complete control over maintaining and supporting it.

Nursing Informatics Applications in Home Healthcare

With the shift to providing extended healthcare at home rather than in a hospital, home healthcare is becoming the new frontier in healthcare informatics. Electronic-based informatics provides healthcare providers with information that has previously only been available in a hospital.

Home healthcare providers use laptop computers and the patient's telephone line or cable connection to connect to patient information stored in hospitals and other offsite locations. Laptops are also used to collection current patient information, which is then electronically sent to the hospital's database application, making this information instantly available to other healthcare providers. This same tool is used by the healthcare provider to receive daily assignments, to process reimbursements, and to communicate directly with administrators and other healthcare providers.

For example, a physician can connect to the hospital's database application from their office and review the home healthcare provider's assessment of the patient before the provider leaves the patient's home. The physician can enter new orders that are immediately made available on the home healthcare provider's laptop.

TELEHEALTH TECHNOLOGY IN THE HOME

Telehealth is a technology that enables a physician to interact with the patient from a remote location. Video and audio equipment and a dedicated computer that measures the patient's condition are placed in the patient's home. Information gathered by this equipment is sent over telephone lines, cable, or a wireless connection to equipment in the physician's office or home. The physician can see and hear the patient—and the patient can see and hear the physician.

The physician can question the patient just as if they were both in the same room. The physician can observe the patient via the video camera. The patient can perform self-assessments, such as blood glucose readings from a glucometer, and blood pressure using an automatic blood pressure cuff. In some cases, a home healthcare provider can be present to assist the physician perform other patient assessments.

As a result of telehealth technology, patients can receive personal, timely care from a physician without the patient having to travel to the physician.

The Informatics Team

An informatics team is a group that represents all the disciplines involved in generating and maintaining information necessary to operate the healthcare facility. Team members include physicians, nurses, unit secretaries, hospital administrators, and the technical staff of the hospital's IT department.

Healthcare providers, hospital administrators, and support staff identify the information that is required to operate the hospital, and work with the IT department to design applications and obtain related computer hardware to acquire this information.

The design process begins when a healthcare provider or hospital administrator recognizes a need. The need could be to acquire new information, analyze existing information, or automate an existing process. A request for proposal is then submitted to the IT department, asking it to come back with a proposal on how to fulfill this need. The IT staff investigates the need by interviewing healthcare providers and others who might shed light on the issue. The proposal returned by the IT department restates the need and proposes a way to meet that need. The proposal also includes resource requirements, a timeline, cost, and a cost-benefit analysis.

Resource requirements identify project resources and ongoing resources. *Project resources* are people and things necessary to fill the need, such as programmers to write an application, and computer hardware to run it. *Ongoing resources* are people and things that are necessary to keep the need fulfilled. Think of project resources as one-time resources that go away once the project is completed. For example, you don't need programmers and the computers used to write the application once the project is completed. Think of ongoing resources as resources that don't go away once the project is completed. For example, the IT department may have to permanently increase its staff with two programmers to support and maintain the new program.

The *timeline* is a schedule of when the project could begin and how long it will take before the need is filled. Included in the timeline are design, development, testing, and implementation of the proposed solution.

The *cost* of the project is broken down into two categories: one-time expenses and ongoing expenses. One-time expenses typically cover everything up to and including implementation of the solution. Ongoing expenses are costs added to the hospital's operating budget.

The *cost-benefit analysis* is the most important piece of the proposal because it tells whether or not the proposal is a good investment. It answers the question, "Will the hospital bring in more revenue or save money by instituting the proposal?"

THE APPROVAL PROCESS

The IT department's proposal is then reviewed by the informatics team to determine if the team should move forward with the proposal. Many hospitals have a formal approval process where various hospital administrators evaluate proposals and decide which are presented to the board of trustees for funding.

Proposals undergo legal and financial review to ensure that they are legally sound and financially wise to undertake. It is also during this review that the funding source for the proposal is identified. Funding can come from a variety of sources, which include the current operating budget, the capital budget, loans from financial institutions, and donations. The funding source must be identified before the proposal is adopted by the board of trustees.

AFTER APPROVAL

Once the board of trustees approves the proposal, the board appoints a project sponsor, who is responsible for the project and makes all decisions related to it. The project sponsor appoints a project manager, who manages the day-to-day operations of the project. The project manager is responsible for assembling the project team, developing the project plan, and delivering the proposed solution on time.

The project manager forms a steering committee, which consists of healthcare providers and hospital administrators. The committee meets once a week with the project manager to assess the project's status and address any issues that the project manager needs to resolve.

The project manager also enlists stakeholders to participate in the project. A *stakeholder* is anyone who is directly or indirectly impacted by the project. For example, a project to create an EMR impacts physicians, nurses, pharmacists, and the unit secretary, among others. These are stakeholders.

A stakeholder's role is to evaluate and advise on the aspect of the project that affects them. For example, pharmacists would evaluate an EMR from their involvement in patient care. The pharmacists' evaluation includes review of the design of data input and the display screens and reports that pertain to the functioning of the pharmacy, along with how the pharmacists want to process the information. They also suggest improvements.

SPECIFICATIONS

With the help of stakeholders, the steering committee, and the project sponsor, the project manager and project team create the specifications for the project. The *specifications* are detailed descriptions of the proposed solutions. This document clearly states how the proposed solutions should work. In the case of an EMR, for example, the specification details data input, data processing, data storage, and data output. Developers use the specification as the blueprint for building and testing the EMR.

Before development begins, the specification is reviewed and approved by the project manager, the steering committee, and the project sponsor. The project is deemed successful if it meets the specification.

USER ACCEPTANCE

Once the project manager is satisfied that the project team has met the specification, the solution is turned over to the project sponsor for the user acceptance test. This is where the project sponsor asks stakeholders to simulate using it in their daily operations. Think of the user acceptance test as test driving a car before purchasing it.

The objective is to determine if the solution fulfills the needs that were identified at the time that the IT department was asked for a request for proposal.

If the solution is accepted, the project manager implements the solution. In the case of an EMR, current patient data is converted for use by the new EMR application and that program is installed on the hospital's computers. The staff is then trained before the EMR application program becomes part of the hospital's standard operation.

If the solution is not accepted, the project manager, with the help of stakeholders, identifies where the solution failed to meet the specifications. The project team then makes the necessary changes, tests it, and begins another round of user acceptance testing.

A common problem that occurs is that stakeholders, in their naiveté and inexperience, expect to see features that were not in the specifications. They think these features are hidden within the written specifications. When they don't see them, stakeholders won't accept the solution. The project manager then explains that the missing features are enhancements, which usually leads to controversy. The outcome may be that the project does not go forward to completion, or the concept of "enhancements" is accepted as something that can be eventually purchased or developed.

Summary

Informatics is the study of information and how information is communicated. Measurement is a process used to analyze empirical data collected by informatics. The combination of informatics and measurement is used to understand what has happened, what is happening, and what is likely to happen in the future.

Informatics is used to devise a system for collecting accurate patient information and disseminating it in the clearest, most useful form to healthcare providers who must care for the patients. Informatics is designed around types of patient encounters, and uses the most efficient and convenient way to collect data and reduce data input errors. Informatics must be designed to ensure the confidentiality of patient information.

Computers are at the heart of informatics technology. A computer is an electronic device that collects, stores, manipulates, and displays information. Computers are categorized as dedicated or nondedicated. Software refers to the set of computer program instructions that tells the computer when to display information on the screen, read data from the keyboard, make decisions, and perform other basic tasks. Computer programs used in informatics and measurement are called applications.

Computer hardware is the physical computer and consists of input devices, storage devices, processing devices, and output devices. A computer network consists of a shared computer, called a server, a router, cables, and other computers (*clients*) used to access the network. Networked services are application programs, databases, and other resources that are available to be shared through the network by client computers.

Home healthcare providers use laptop computers and the patient's telephone line or cable connection to connect to patient information stored in the hospital's database and other offsite locations.

Telehealth enables a physician to interact with the patient from a remote location. Video and audio equipment and a dedicated computer that measures the patient's condition are placed in a convenient place in the patient's home.

An informatics team is a group that represents all the disciplines involved in generating and maintaining information necessary to operate the healthcare facility. A project sponsor is responsible for new projects and makes all decisions related to the project. A project manager manages the day-to-day operations of the project. A steering committee consists of healthcare providers and hospital administrators who meet weekly to assess the project's status and address any issues that need to be resolved. A stakeholder is anyone who is directly or indirectly impacted by the project.

The specification is a detailed description of the proposed solution. It clearly states how the proposed solution should work and is used to determine whether the project has achieved its goal.

Quiz

1. The study of information and how information is communicated is called:

    (a) Informatics

    (b) Measurement

    (c) Electronic information

    (d) None of the above

2. An EKG monitor is an example of:

    (a) Nonelectronic informatics

    (b) A nondedicated computer

    (c) A dedicated computer

    (d) None of the above

3. Microsoft Word is an example of:

    (a) A commercial application program

    (b) A customized application program

(c) An informatics application

(d) None of the above

4. A router redirects requests to the appropriate computer on the network.

(a) True

(b) False

5. Reasonableness testing is performed to ensure that data makes sense.

(a) True

(b) False

6. What does EMR stand for?

(a) Employee Medical Record

(b) Electronic Medical Record

(c) Electronic Medication Record

(d) None of the above

7. A user acceptance test is used by the project sponsor to learn if the project meets specifications.

(a) True

(b) False

8. Technology enabling a physician to interact with the patient from a remote location is called:

(a) Telephony

(b) Telehealth

(c) Telemedicine

(d) All of the above

9. Building an in-house application program is expensive because the hospital underwrites the entire expense.

(a) True

(b) False

10. The healthcare provider can verify patient provided data with data supplied by the healthcare provider that is known to be valid.

(a) True

(b) False

CHAPTER 13

Risk Management

Are you a risk taker? You may not think so, but everyone is a risk taker to some extent because nothing is certain except for death and taxes. Even if you stay in bed, you're risking getting a decubitus that could lead to your demise.

Some people are better risk takers than others, because they know how to manage risk by reducing their exposure to situations where the risks outweigh the benefits. A Nurse Manager must become a smart risk taker to care for patients and the staff.

This chapter takes a close look at the risks involved with healthcare and how to manage those risks.

What Is Risk Management?

Risk management is the process of evaluating situations that might lead to a loss and then selecting a course of action to eliminate the risk, reduce the likelihood that the loss will occur, or reduce the impact should the loss occur.

You use risk management daily without realizing it by buckling up your seat belt before driving your car, driving at a reasonable speed, and adhering to traffic signals.

These actions won't eliminate the risk of injury if you collide with another vehicle, but they reduce the chances you'll cause the accident, and minimize your injuries should one occur.

Healthcare is prone to many risks, because healthcare is both an art and a science. A healthcare provider can use science to restore a patient's health only after the patient's condition is properly diagnosed. Diagnosing a patient is an art honed by years of experience by the healthcare provider, and usually inexperience by the patient. With both parties focusing on the best outcome for the patient, it nevertheless can lead to a situation involving risk and ultimately harm to the patient.

When a patient presents to a healthcare provider, both the patient and the healthcare provider have the same goal—treat the illness or improve the patient's own systems so that his or her body may restore itself and the patient can return to the activities of daily life. However, neither the patient nor the healthcare provider is certain of the outcome. Neither is sure that the healthcare provider can restore the patient's health.

The patient, when seeking treatment, is risking his health on the healthcare provider's skills and experience. The patient manages this risk by choosing a healthcare provider who has a proven record of successfully caring for patients—and by reading about symptoms, medical tests, diagnoses, and treatments.

Suppose a patient presents with a dry cough that has been nagging him for eight weeks. After an examination, the healthcare provider concludes that the patient has allergies, because his lungs are clear and it is a very bad allergy season. The patient says he's never had allergies. The healthcare provider responds by saying "our body changes as we get older." The patient questions this conclusion and knows that ignoring the cough is too risky, because a persistent dry cough is one of the first symptoms of lung cancer. The patient insists on a chest x-ray as a way to manage this risk.

The healthcare provider risks their assets whenever they care for a patient. Each time a patient presents, the healthcare provider is exposed to the risk of losing their home, possessions, livelihood, and health. The healthcare provider manages these risks by knowing their limitations, adhering to generally acceptable medical procedures, and acquiring malpractice insurance.

When presented with a patient who has a persistent cough and no history of allergies, the healthcare provider mitigates the risk of reaching a wrong diagnosis by ordering a chest x-ray. The healthcare provider further lowers their risk by asking a radiologist, who spends all day interpreting x-rays, to read the x-ray. These actions don't prevent an erroneous diagnosis, but they do go a long way toward minimizing its likelihood.

Healthcare facilities are also exposed to situations that place patients, healthcare providers, staff, visitors, and the institution itself at risk. A healthcare facility is exposed to greater risks than a healthcare provider because of the volume of patients who are being treated, the severity of their illnesses, and the complications of their treatments.

Managing Risk

The initial step to managing risk is to conduct a risk audit to identify previous losses within the healthcare facility—as well as in other healthcare facilities—and determine how losses occurred. A risk audit is an ongoing process performed by the Nurse Manager in conjunction with the healthcare facility's risk management department.

A risk audit is triggered by the filing of an *incident report*, a written documentation of an event that might have deviated from the healthcare facility's policy or could lead to a loss. For example, suppose that a patient was given 5 mg of Coumadin when the order called for 10 mg. Not receiving the proper amount of the drug could lead to vascular problems for the patient, and perhaps a setback in their recovery. Upon discovering the medication error, the Nurse Manager should make sure that the patient's nurse files an incident report with the risk management department. It is also the Nurse Manager's responsibility to find out why this happened, and how the likelihood of this recurring can be reduced.

A risk audit is also triggered by a claim filed by the patient. A *claim* is an oral or written complaint alleging mistreatment by the healthcare facility. The claim can take several forms. These include a comment made to the Nurse Manager upon discharge, a response to a patient survey, a letter written by the patient to the healthcare facility's administrator, or legal action taken by the patient's attorney.

Finally, a risk audit can be triggered by an administrator's or manager's request to evaluate a new product, new equipment, or a new procedure to ensure that it is safe for use in the healthcare facility.

ANALYZING RISK

Once a risk audit is triggered, auditors from the risk management department piece together everything that led up to the event. The goal is not to affix blame, but rather to understand what happened and how to prevent it from happening in the future.

An auditor is likely to be an RN and possibly a Nurse Manager who has the necessary medical training to investigate a situation. The auditor begins by assembling facts. The type of facts acquired depends on the nature of the audit.

If the incident involves a patient, the auditor needs to learn about the patient, the staff that cared for the patient, and how the unit operates by reviewing the patient's records, staffing records for the unit, and qualifications and employment records of each staff member who was directly or indirectly involved in the incident.

Next, the auditor determines what should have happened. What should have been the care given to the patient based on the patient's diagnosis, treatment plan, medical orders, and the policies and procedures that relate to the care of this patient?

Next, the auditor determines what did happen. That is, what care did the patient receive? The auditor makes this determination by reviewing the patient's records and the incident report. The auditor may ask the Nurse Manager, the Nurse Manager's staff, and the staff of other departments who are involved in the patient's care to provide oral and written statements describing the incident.

An oral statement is an interview in which the auditor asks questions and notes the staff member's responses. The auditor may follow up with additional questions. At the end of the interview, the auditor typically sums up the responses to ensure there isn't any misunderstanding. A written statement is an addendum to the original incident report in which the auditor asks the staff member to describe the incident, and may ask the staff member to respond to specific questions in writing.

The auditor then draws a conclusion that there either was or wasn't an actual or potential loss. If there was an actual or potential loss, the auditor identifies factors that led to it and suggests changes to avoid it occurring in the future; such as changes to policies and procedures or staff retraining. In extreme cases, the matter might be turned over to the HR or legal department for action.

If the audit is to evaluate a new product, equipment, or procedure, the auditor must establish acceptable criteria that can be used to objectively measure the risk of using the new vs. current product, equipment, or a procedure. This is accomplished by review of the literature on the safety of the product or equipment, and seeking best practice examples from other, similar organizations, regarding a procedure.

The auditor then compares the current risks to the potential risk of changing to the new product, equipment, or procedure and determines which is less risky.

REDUCING EXPOSURE TO RISK

A key role of the risk management department is to lead a coordinated effort by the staff, management, and hospital administrators to effectively reduce the likelihood that a loss will occur resulting in injury to a patient or a staff member. This effort focuses on the following:

- **Staff training** The staff needs to be shown how to care for a patient using less risky procedures. Having potential risk protocols is also helpful. For example, a "fall risk checklist" should be filled out on every newly admitted patient. Each factor is given an objective score. Factors include the patient's age, previous history of falling, history of stroke or syncopy (losing consciousness), and any signs of disorientation. The patient is given a score and, if high enough, is put on fall precautions. These precautions may include requesting that the patient ask for assistance when going to the bathroom, or referring the patient to physical therapy for instruction on innovative ambulation techniques.

- **Adequate staffing**   The Nurse Manager and hospital administrators must provide adequate staffing levels to ensure that the staff has sufficient time to deliver patient care.

- **Consistent performance**   A standard level of performance must be clearly defined for everyone involved in patient care.

- **Performance measurement**   Department managers must evaluate the performance of healthcare providers, including physicians, to ensure that performance standards are maintained.

- **Disciplinary action**   Disciplinary policies must be established and strictly enforced to ensure that healthcare providers who don't meet performance standards are suspended from delivering patient care until they receive remedial training.

- **Misinterpretation**   Eliminate situations where medical orders must be interpreted by nurses and other healthcare providers. All orders should be clearly written without using abbreviations.

- **Contraindications**   All potential medication contraindications should be clearly identified on medication packaging and medication reports.

- **Drug incompatibilities**   All drug incompatibilities should be identified and resolved by the pharmacy before medication is delivered to the unit.

- **Safety standards**   Record the number of decubitus, nosocomial infections, iatrogenic actions, and other conditions that a patient contracts in a hospital. Also monitor conditions that lead to patient falls.

- **Medication errors**   Procedures must be in place to make sure that the patient receives the right medication, at the right dose, at the right time, and using the right route.

- **Proper documentation**   Care must be documented immediately after it is given to the patient.

- **Carry out orders**   All medical orders must be carried out correctly and on time.

- **Patient assessment**   Physicians, the Nurse Manager, and other healthcare providers caring for the patient must be notified of changes in a patient's condition immediately.

- **Environmental hazards**   Environmental hazards must be reported and immediately removed.

- **Equipment malfunctions**   Equipment that requires repair must be taken out of service immediately.

The Incident Report

An incident report, sometimes referred to as a sentinel event report, is a permanent, written statement of the event, and is used by the risk management department to prevent incidents and to refresh memories of the event should it result in legal action. An incident report is written by the staff member who is most closely involved with the incident, such as the person who first becomes aware of it.

The Nurse Manager must make sure that an incident report is filed with the risk management department every time that a risky event occurs on their unit. In this way the event can be investigated and steps can be taken to prevent it from occurring again, not only on the Nurse Manager's unit, but on units throughout the healthcare facility.

Each healthcare facility has its own style of incident report; however, most incident reports contain the following type of information:

- **Date and time the incident report was written**   The staff member involved in the incident writes a description of the incident as soon as possible after it occurs.
- **Date and time of the incident**   This is the date and time of the actual incident.
- **Name of the unit**   The unit where the incident occurred.
- **Location**   The location of the incident.
- **Staff member**   The name and other information (e.g. unit, employee number, and so forth) that identifies the person who is writing the incident report.
- **The description of the incident**   This section of the incident report is where the staff member describes the incident in their own words, stating only facts, not hypotheses.
- **Signature**   The staff member must sign the incident report.

The description of the incident must be explicit and answer the questions who, what, where, where, and how about the incident. The patient and any staff members must be identified by their formal name, using any identifying numbers if available. For example, Mary Jones, RN charge nurse, is preferred over Mary, the patient's nurse. Likewise, Mr. Robert Smith, patient number 12345, room W203-1, is better than saying Robert Smith the patient, because the former clearly identifies the question of who is involved in the incident.

The narrative should flow chronologically, starting with the moment that the staff member who is writing the incident report became involved in the incident. The staff member might say, for example "On or about 11:30 A.M. I administered 1 unit of regular insulin to Mrs. Betty Right (patient number 4325) in room W202-2. I left her room at 11:40 A.M. and noticed Mr. Tom Paterson (patient 8768) lying on the floor in room W204-1."

Stating the time along with your action is very important in an incident report, because it enables the risk management department to reconstruct your activities along with the activities of others involved in the incident. Do not draw conclusions or elaborate or embellish the facts.

Don't draw a conclusion if you didn't see the incident actually happen. It is common to come upon the scene after the incident occurs. You see certain things and probably make assumptions about things you don't see. Only describe things that you witness. Let's say a new nurse to the unit is on medication rounds. She's in the room alone with the patient and calls a code. You're the first responder and see a syringe on the table and the nurse performing CPR. It is easy to suspect that the nurse gave the patient inappropriate medication. However, this suspicion should not be in your incident report because you don't know this to be true. You could write that when you arrived at the code, a syringe was on the patient's table and Mary Salle, RN was performing CPR.

Likewise, if you see a patient lying on the floor, you can't write that the patient fell out of bed, because you didn't see the patient fall. You can say, "I entered the patient's room and discovered the patient lying supine on the floor. Both side rails on the bed were raised. The patient was unresponsive to my calling her name."

Here are some examples of circumstances in which you need to write an incident report:

- A patient is injured
- Unanticipated death
- Adverse drug reaction
- Malfunctioning equipment
- Violent behavior from anyone on the unit
- Smoke, fire, toxic spill, or any environmental emergency

THE BLAME GAME

While the risk management department and hospital administrators view an incident report as a way to improve staff and patient safety, staff might see the incident report as an admission of guilt, because it might highlight their errors and have far-reaching personal repercussions. This is especially true in our litigious society where words contained in an incident report can be used against the staff in court.

Everyone makes honest mistakes—and most people admit to them. However, few people will tell the whole truth if it exposes them to personal hardship. Instead, they will likely tell half-truths rather than outright lies, because being caught in a lie also has serious personal consequences. Informal, verbal communications may involve such half-truths, but an incident report notes the facts of the situation, not opinion, and therefore is not very likely to contain misinformation.

This is not true in my experience and was inadvertently included in this chapter. This is also not true.

Common Risk Exposures

Although each patient's course of treatment opens the patient, the staff, and the healthcare facility to their own individual types of risks, these risks can be generally grouped together in several classifications:

- **Patient relations**   Patients expect to be treated with respect and kindness by all members of the healthcare team. Patients also want to be made part of decisions about their health. When these expectations don't materialize, patients become disgruntled.

- **Outcome of treatment**   When the treatment or procedure fails, the patient is likely to blame the healthcare team. Even when it succeeds, the healthcare team could be blamed for any unpleasant side effects. This happens even when the patient signs an informed consent.

- **Lack of policy**   Healthcare facility policies describe how to care for patients while exposing the patient, staff, and healthcare facility to the minimum amount of risk. However, at times these policies become outdated, due to new treatment protocols or newly acquired pieces of equipment.

- **Staff behavior**   The staff is expected to maintain the same standard of care, regardless of the patient's condition and behavior. Incidents occur when these standards are not met.

Studies have shown that untoward incidents are related to procedural or systems issues, rather than the actions of a staff member. For example, malfunctioning equipment may not be taken out of service and repaired, because there is an inadequate routine system of assuring periodic checks and/or removal of malfunctioning equipment. Similarly, existing policies may not be enforced, enabling some staff members to avoid them, thereby exposing themselves and the patients to unnecessary risk.

The Risk Management Department

The risk management department, usually a subunit of the healthcare facility's legal department, oversees insurance contracts between the healthcare facility and insurers as well as managing claims. The department staff works hand-in-hand with insurers to minimize exposure to situations that have a high probability of resulting in an insurance claim. This lowers the number of claims and, as a result, lowers premiums for the healthcare facility.

Many insurers provide healthcare facilities with in-service courses that show the staff throughout the facility how to avoid risky situations. Insurers also offer healthcare administrators risk management services where consultants review the healthcare facility's operation, looking for practices that expose the healthcare facility to higher claim rates.

The risk management department also has its own staff of auditors who examine claims, review and analyze incident reports, conduct risk management surveys, and develop policies and procedures that reduce risk to patients, the staff, and the healthcare facility.

In many healthcare facilities, the risk management department is also responsible for ensuring that the staff and the healthcare facility are compliant with federal and state laws and requirements of the Joint Commission on Accreditation of Healthcare Organizations (JCAHO).

The Nurse Manager and Managing Risk

Nurse Managers play a critical role in managing risk because they are on the front lines of healthcare where risk incidents are most likely to happen. The Nurse Manager is responsible for reducing risky situations by ensuring that staff members follow the healthcare facility's policies and procedures.

When an incident involving a patient occurs, the Nurse Manager may be the first on the scene and must take remedial action to minimize any imminent danger to the patient and staff. The Nurse Manager then conducts a preliminary investigation to determine what happened, and why it happened, before notifying the risk management department, which then conducts a formal inquiry. If the Nurse Manager is unavailable, the person in charge of the unit at the time of the incident will report it.

During the preliminary investigation, the Nurse Manager determines if care was given within the scope of the Nurse Practice Act and the standards of care for the patient's condition. If so, then these facts mitigate a claim of professional negligence. The preliminary investigation should include the following:

- Private interviews with the patient and each staff member who is directly or indirectly involved in the incident to determine their perceptions of the facts surrounding the incident.
- Review of the patient's chart to determine if it is timely and accurate.
- A determination of whether all medical orders were carried out on time and according to the standard of care.
- Examination of nurses' notes to ensure that they are up to date.

Once the preliminary investigation is completed, the Nurse Manager must ensure that the incident report is written. The incident report should be written by the staff member who was closest to the incident. Other staff members involved in the incident can add their own addendums to the incident report. Depending on the policy of the healthcare facility, the Nurse Manager may not have to participate in the incident report unless they too were involved in the incident. The Nurse Manager's primary role is to make sure that the appropriate staff member writes the incident report.

The incident report must be kept separate from the patient's medical record. The patient's medical record, including nurses' notes, is usually turned over to the patient's lawyer as part of the discovery process should the incident lead to litigation. However, an incident report is not part of the discovery process in many states, and thus should not be part of the medical record.

Attorneys for the healthcare facility and the patient will carefully examine the patient's medical records, looking for inconsistencies and gaps that can be construed as professional negligence. *Professional negligence* is failure to administer care that a reasonable nurse would render under the same or similar circumstances.

REDUCING INFORMATICS RISKS

Computerized informatics reduce some risks that are common in the unit, such as misinterpreting medical orders, overlooking contraindications before administering medication, and failing to carry out medical orders on time. This is because computers can be programmed to either avoid those errors (e.g. misinterpreting medical orders), or display a warning message when an error is detected (overlooking contraindications). Computers can also be programmed to print a schedule for each nurse containing scheduled treatments for the nurse's patients.

However, computerized informatics expose the patient, staff, and healthcare facility to new risks, many of which can be easily mitigated by the Nurse Manager and staff by adhering to the following:

- **Protect passwords**   Each staff member creates their own password giving them access to confidential medical records. The password must remain confidential and not be shared or written down in a location that is easily accessed by anyone else.

- **Log off**   After accessing medical records, log off the computer to eliminate the risk that someone else will use your login to continue accessing medical records.

- **Erase the computer screen**   Don't leave a patient's medical record on the computer screen once you access the information necessary to care for the patient.

- **Conceal the computer screen**   When viewing a patient's medical record, make sure that the computer screen is viewable only by authorized staff.

- **Correct mistakes immediately**   Each staff member must learn how to correct any data entry errors and ensure this is done when errors occur.

REDUCING PERSONAL RISK

The Nurse Manager and staff members are at personal risk when caring for patients. Any inadvertent error could lead to litigation and the loss of personal assets and the license to practice nursing. While this is an extremely rare occurrence, many nurses have personal malpractice insurance to guard against this circumstance.

As a Nurse Manager, the best way to manage personal risk is to make sure that you and your staff adhere to the standard of care, especially for the patient population of your unit. Standards of care may differ depending on the patient's condition.

Be sure to stay abreast of the latest clinical procedures by attending in-service courses and outside conferences, keeping up-to-date on literature, and becoming certified in the specialties of your unit.

All members of the staff should have their own professional liability insurance, which is available for a reasonable premium from nurses' associations and private insurance companies. Don't rely on the healthcare facility's umbrella professional liability that covers its nursing staff, because such policies are designed to protect the healthcare facility and do not necessarily protect the nurses' assets.

Take particular precautions against exposure to fringe litigation that might arise even if there is no apparent incident. For example, a disgruntled patient may take legal action if a nurse didn't show concern for the patient and the patient's family, or if the patient feels that the nurse failed to provide adequate information about medication or treatment. Although such litigation may be without merit, you still have to defend yourself by hiring an attorney.

Here are tips to reduce this exposure:

- Go out of your way to show concern for patients and their families.
- Have the patient participate in developing the patient's care plan.
- Explain the reason that the patient is receiving medication and treatment. Be sure that the patient acknowledges their understanding of your explanation. Give the patient literature about the medication or treatment, if available.
- Scan the patient's room for anything that exposes the patient to the risk of injury, and then rectify the situation.
- Delegate patient care wisely. Remember that you can be sued if the delegated task is performed improperly.

REDUCING RISK USING PATIENT CONSENT AND DIRECTIVES

A patient shares the risk of medical care. However, it is the responsibility of the healthcare provider to identify those risks and have the patient sign an informed consent before beginning treatment.

An *informed consent* is a document that must clearly state in laymen's terms the nature of the treatment or procedure, its benefits, and possible side effects, including disability and death. It could also include alternative treatments that were considered, and the cost of the treatment or procedure, depending on the healthcare facility's policy. There are many kinds of informed consent, including those for emergencies, for experimental treatment, and for care of a minor.

The primary healthcare provider—not the nurse—is responsible for explaining the treatment or procedure. The nurse can provide further explanation to the patient once the healthcare provider finishes their explanation.

The nurse must ensure that the informed consent is signed before beginning the treatment or procedure. The nurse takes on the role of patient advocate during this process, and must make sure that the patient is capable of understanding the contents of the informed consent, does understand it, and is signing it voluntarily. The nurse should ask the patient to explain the contents of the informed consent before it is signed. If the explanation isn't adequate, then the nurse should withhold the informed consent and discuss the situation with the primary healthcare provider, who will likely explain the treatment or procedure to the patient again.

In many healthcare facilities, the nurse is asked to sign the informed consent after the patient signs it. The nurse's signature is as a witness to the patient's signature.

A patient specifies the type of care they should receive if they become unconscious by signing a *directive*. There are three commonly used directives: a do not resuscitate (DNR), a living will, and an advance directive.

A DNR is a directive that states the patient does not want to be resuscitated should they go into cardiac arrest and respiratory failure. Instead, the patient should be kept pain-free and let nature take its course. A DNR directive must be placed in the patient's chart. A DNR can be revoked at any time by the patient simply by telling the nurse or any staff member that they no longer want the DNR directive.

If the patient doesn't have a DNR and wants one, the hospital will help them write one. A difficult situation occurs when a patient who has a DNR becomes unconscious and family members oppose the DNR. Although the patient's request should be followed, it is best to send the issue to the hospital's ethics committee, which will then decide the proper course of action.

An advance directive, which also can take the form of a living will, comprises instructions written in advance of becoming ill or incapacitated. This document states the patient's wishes regarding medical treatment. The contents may include instructions for any procedure or treatment, such as not to be put on a respirator, not to receive kidney dialysis treatment, not to be resuscitated if breathing stops, and not to be given a feeding tube.

Another type of advance directive is a durable power of attorney for healthcare, which names a person to make decisions for the patient if the patient is unable to express their own decision regarding healthcare.

Summary

Risk management is the process of evaluating situations that might lead to a loss, and then selecting a course of action to eliminate the risk, reduce the likelihood that the loss will occur, or reduce the impact should the loss occur.

Risks are managed by conducting a risk audit to identify previous losses within the healthcare facility—as well as in other healthcare facilities—and determine how losses occurred. A risk audit is triggered by the filing of an incident report, written documentation of an event that might have deviated from the healthcare facility's policy or could lead to a loss. A risk audit is also triggered by a claim filed by the patient. Finally, a risk audit can be triggered by an administrator's or manager's request to evaluate a new product, equipment, or procedure to ensure that it is safe to use in the healthcare facility.

The risk management department analyzes the result of a risk audit to understand what happened and how to prevent it from happening in the future. The risk management department is the leader in coordinating efforts by the entire healthcare facility to reduce the risk of a loss to the patient, staff, or healthcare facility.

An incident report, sometimes referred to as a sentinel event report, is a permanent, written statement of the event and is used by the risk management department to prevent incidents, as well as to refresh memories of the event should it result in legal action.

The Nurse Manager plays a key role in managing risk and is responsible for reducing risky situations by ensuring that the staff follows the healthcare facility's policies and procedures. When an incident occurs, the Nurse Manager first takes remedial action to minimize the danger to the patient and staff, and then conducts a preliminary investigation to determine what happened and why it happened.

Quiz

1. What step reduces risk to patients, staff, and the healthcare facility?

   (a) Informatics

   (b) Staff training

   (c) Performance measurement

   (d) All of the above

2. The description of an incident in an incident report should:

   (a) Describe in clear terms what the nurse thinks happened

   (b) Flow chronologically

   (c) Contain the opinion of the Nurse Manager

   (d) All of the above

3. The healthcare facility's insurer helps to reduce risk by:

   (a) Providing in-service training

   (b) Conducting risk surveys

   (c) Analyzing claims

   (d) All of the above

4. An incident report is part of the patient's medical record.

   (a) True

   (b) False

5. A Nurse Manager and staff members should obtain their own professional liability insurance.

   (a) True

   (b) False

6. The risk management department is responsible for:

   (a) Reducing exposure to risk

   (b) Managing claims

   (c) Overseeing insurance contracts for the healthcare facility

   (d) All of the above

7. Computerized informatics is risk-free.

   (a) True

   (b) False

8. A DNR:

   (a) Can be rescinded at any time

   (b) States the patient's wishes for the use of extraordinary life-saving procedures

   (c) Is governed by state law

   (d) All of the above

9. A nurse is the patient advocate when a patient is signing an informed consent.

   (a) True

   (b) False

10. In many states, an incident report is not part of the discovery process in litigation.

    (a) True

    (b) False

CHAPTER 14

Managing Scarce Resources

"I have too few registered nurses and not enough money in the budget to give my patients the high-quality healthcare that they deserve." Nurse Managers are heard saying this every day to hospital administrators...only to hear, "Yes, but you'll have to live with it."

Money and RNs are just two of many scarce resources needed to care for patients. The problem: there aren't enough resources to go around. The solution: manage scarce resources so that patients continue to receive a high level of care.

In this chapter, you'll learn techniques to manage both scarce human and nonhuman resources.

What Is a Scarce Resource?

A *resource* is something needed to complete a task, and a *scarce resource* is a resource that has limited availability because there is more demand for the resource than availability of the resource.

A Nurse Manager is responsible for nursing patients back to good health by using resources provided by the healthcare facility. The most important resource is a budget, which is that amount of money required to acquire other resources such as staff members, equipment, and supplies.

In an ideal world, a Nurse Manager has sufficient budgetary resources to give patients the best care that money can buy. However, reality quickly sets in when there are insufficient resources to meet this idyllic goal. This means the Nurse Manager must use available resources to care for patients. Many of the available resources are in demand by other departments within the healthcare facility and by competing healthcare facilities, thus making these resources scarce. The Nurse Manager must develop skills to deliver healthcare by managing the best use of these scarce resources.

THE ECONOMICS OF SCARCE RESOURCES

Although family members of patients are frequently heard telling a physician to spare no expense (they don't really know what that means) when treating their loved one, there are economic considerations that limit patient care. Few patients have unlimited financial resources to receive unlimited healthcare, especially in cases where returning to activities of normal daily life is a remote possibility.

Financing healthcare is the initial scarce resource that must be managed. Putting aside the emotional desire of family members, physicians, Nurse Managers, and other healthcare providers must provide treatment that is reimbursed by a third party, such as an insurer or government agency, based on a complex set of regulations designed to make healthcare affordable.

With help from hospital administrators, the Nurse Manager can manage scarce financial resources by keeping these regulations in mind when caring for patients. For example, some insurers require pre-approval before certain tests are performed on a patient; otherwise, the healthcare facility will not be reimbursed for the test.

As a Nurse Manager, be sure to instill in your staff a sense of fiscal responsibility. Each of your staff members should have a general idea of the cost of medication and other supplies used in the care of patients. They should also have a clear understanding of how the healthcare facility is—and is not—reimbursed for these expenses. For example, the cost of medication wasted in error is not reimbursed and must be borne by the hospital.

Your staff should have a working knowledge of the rules of various healthcare plans, such as managed care plans, Medicare, and Medicaid. This gives them a full picture of how their patient's healthcare—and the healthcare facility—are financed, and an appreciation of decisions to limit resources to the unit.

Explain to your staff the difference between the price the hospital charges patients and third-party payers, and the actual cost incurred to care for the patient. The price covers patient care plus overhead and an adjustment to reflect reduced prices charged by third-party payers.

Contrary to popular belief, all patients are not guaranteed the same level of medical care. They are guaranteed a level of care that they can afford, either directly out-of-pocket or through third-party medical coverage.

MANAGING SCARCE FINANCIAL RESOURCES

As a Nurse Manager, your unit's budget is a scarce resource that you must manage to ensure that all primary and secondary needs of your patients, your staff, yourself, and your unit are addressed. A *primary need* is a resource that is required to achieve your primary goal, such as having sufficient nurses for each shift, and having the equipment and supplies they need to care for patients. A *secondary need* is a resource that indirectly supports a primary need, such as in-service training to maintain a high level of staff competence when caring for patients.

Your preliminary budget contains funds for your unit's primary and secondary needs. You must develop a political strategy to ensure that all funds remain in the approved budget, and are not later removed during the year as adjustments are made to the budgets to keep the hospital within its overall fiscal goal.

Your job is to develop a strong, unshakable rationale for each item in your budget and relate it to patient care and reimbursements. The following are ways to achieve this:

- **Benchmark other hospitals**   Compare your staffing ratios and other key indicators with those of comparable healthcare facilities to justify your preliminary budget.

- **Centralize staffing**   Demonstrate how you adequately utilize the services of a centralized staff office that assigns per diem nurses to units to balance staffing between full-time and per diem personnel.

- **Reduce staff turnover**   Show how your preliminary budget includes items such as staff training that increase staff moral and reduce turnover—savings the $40,000 expense of replacing a nurse. The cost is higher when training a nurse in a more specialized area, such as critical care, ER, or OR.

Campaigning is critical to getting your preliminary budget adopted—and having it remain intact during the fiscal year. Cultivate personal connections to your supervisor and administrators who approve your budget. At opportune times, mention your unit's successes and challenges and how you built into your budget funding for meeting those challenges in a fiscally reasonable way. Be persistent in your campaign, but not obnoxious.

Leverage friendships you might have with board members and political leaders in communities serviced by your hospital. Tactfully discuss situations the hospital faces, and how hospital administrators can handle them with support from the board and community leaders. Always be supportive of management in your conversations and avoid anything that might imply you are going over their heads to gain support.

Remember that your goal is to be a champion for your unit and not a divisive factor for the hospital. Don't pit your unit against other units. Instead, focus on how your preliminary budget will economically solve one of your hospital's major problems.

Look for creative ways to fulfill secondary needs by using resources from outside your budget. For example, formula and pharmaceutical firms are funding sources for in-service training programs. Your budget still must cover your staff's time, but cost for the trainer and training supplies is funded outside your budget.

Train staff for positions that are difficult to fill. Let's say that you manage your hospital's critical care unit. Increase your training budget so that you can offer training to staff from other units who show an interest in your unit. Encourage the hospital to have a float pool of nurses who can work in a variety of areas as the need arises on different, but similar, units.

INFLUENCING THE ALLOCATION OF SCARCE RESOURCES

As a Nurse Manager you have power within the healthcare facility to influence how scarce resources are allocated. Your power comes not only from your position as a manager within the organization, but also from your personality traits, called *personal power*.

Think for a moment. Who do you go to when you want a piece of equipment on your unit repaired immediately? Chances are you cut through your organization's red tape by calling your technician friend in engineering. Somehow the technician manages to squeeze in your request either before or right after the next one in the queue. This happens not because you are the Nurse Manager of the unit, but probably because of your charisma, your way with words, and the ongoing respect and fairness you use when interacting with the staff of other departments in the hospital.

Personal power also comes from expertise that others respect. For example, you may have been in a situation where a resident is performing a procedure for the second time in her career. She is in charge, but you lead her through the procedure by asking a series of questions, such as, "Doctor, do you want to use a butterfly bandage instead of sutures?" In this instance, you influence the outcome of the procedure without being in authority.

Information is another source of power you can use to influence allocation of scarce resources. Hospital administrators and managers who are removed from the day-to-day operations base their decisions on perceptions derived from data that has been interpreted for them in reports. Reports give an objective view of the operations only if you assume that the raw data and subsequent analysis accurately portray what is happening on the unit. However, reports are secondary information to the personal knowledge of the Nurse Manager regarding the unit's operation. To the Nurse Manager, a patient is a person and not a number on a report. The Nurse Manager knows about problems that impact patient care but that may not find their way into a report. This is information that upper management and hospital administrators don't have that the Nurse Manager can use to influence their decisions when allocating scarce resources.

For example, hospital administrators tend to generalize when looking at nurse-to-patient ratios. Although they realize that patients require different levels of care, they overlook the skill levels of nurses on a unit. As a result, they assume all nurses have relatively the same skills, which is not necessarily the case. A ten-year veteran of the medical-surgical unit will likely find it challenging to spend a shift in a busy ER.

As a Nurse Manager, use information that you have gathered as a foundation to support your argument for resources. Summarize typical cases that will persuade higher management to reexamine decisions that are based on statistical reports and consider your information as part of the decision process.

Causes of Scarce Resources

Second only to scarce financial resources is the shortage of registered nurses. The average RN is 45 years old, an age at which the physical demands of bedside care are challenging, forcing many to seek less-taxing healthcare roles.

Although nursing schools have seen a dramatic increase in applicants, only a relatively few are accepted because of too few faculty members. It is projected that half the current faculty members will retire by 2010. And too few RNs are pursuing the masters and doctoral degrees required to join a nursing school's faculty. As a result, there are not enough nursing instructors today to teach all those who want to join the nursing profession—and there will be even fewer instructors in the near future.

A Nurse Manager must be creative to manage the unit's nursing demands effectively. Here are a few tips on how, as a Nurse Manager, you can manage scarce personnel:

- **Estimate demand daily**   Based on your unit's census and a nursing care delivery model (see Chapter 3), you can accurately project the number of RNs and supporting staff needed to care for your current patients.

- **Compare needs to staff competency**   Personal knowledge of your staff members' capabilities and information provided by the HR department and colleagues enable you to match your staffing needs to the nurses and support staff scheduled for each shift.

- **Develop a contingency plan**   Develop an ongoing arrangement with sources of temporary staffing so that you can call upon them when your staff is unable to adequately care for your patients. Sources include per diem nurses, a float pool, temporary (i.e. shift) transfers from other departments, and outside staffing agencies.

- **Build a flexible scheduling system**   Less stringent work schedules provide flexibility to match nurses with periods of high demand for patient care.

Defend your staffing needs. Nurse Managers throughout the healthcare facility face the same challenges of managing a limited number of RNs. Hospital administrators typically react by changing the skill mix of a unit's staff, leaning more toward non-RNs providing patient care. Rebuff such attempts by demonstrating how lowering your present staffing levels places your patients' health in jeopardy and the hospital at risk for litigation.

Develop a solid argument for retaining existing staffing levels by gathering facts to prove your position. Note the number of admissions, discharges, and transfers that occur daily. These tasks are time consuming and must be performed by an RN.

List special skills needed to care for your average patient. These should include languages spoken and expertise in specialty areas, such as ER and CCU. Remind the administrator that special-needs patients require care from nurses who have special skills.

Consider the layout of your unit, such as the distance a nurse must walk between rooms, and from the nurses' station to each room. Nurses require more time per patient if they must walk along long corridors.

Identify the type and quantity of technology available in each room. Particularly mention the maintenance record of equipment, and how equipment downtime impacts your nurses. Administrators must not assume that all time-saving equipment on your unit is operational.

Managing Scarce Resources Using Technology

Technology offers opportunities to reduce time-consuming tasks that interfere with the primary responsibility of allocating scarce resources. It is also used to efficiently manage the staff so that each staff member is fully utilized based on their skills. This is especially useful in rural areas.

Most decisions about technology are made by middle and upper management based on long-range strategic goals of the healthcare facility, because technological changes impact departments throughout the organization. However, the Nurse Manager can and should initiate discussions about technological upgrades by demonstrating how investment in technology can better utilize scarce resources while improving patient care and reducing liability risks for the hospital.

Here are ways in which specific types of technology can be used to manage scarce resources:

- **Intercom**   An intercom linking the nurses' station with each patient's room reduces the nurse's response time to assist a patient. The nurse can discuss with the patient their needs without leaving the nurses' station, and then gather necessary supplies before visiting the room to care for the patient. This is an improvement over the patient's call button, because the call button requires the nurse to make two unscheduled trips to the patient's room—one to assess the call, and the other after gathering supplies.

- **Telemetry**   Transducers such as EKG leads can be used to take ongoing physical measurements of the patient that are transmitted to monitors in the nurses' station. This enables the nurses to monitor all their patients at a glance, and reduces unscheduled visits to the patients' rooms.

- **Analytical computer software**   Computer software can analyze incoming telemetry and display the patient's condition, such as heart rate and heart rhythm, on monitors at the nurses' station. It can also detect atypical conditions, such as sinus tachycardia, and alert the nurse. Analytical computer software is also used in electronic charting to detect potential errors and take action to prevent the errors from occurring. It can warn the nurse of contraindications for prescribed medications—and determine if those conditions exist by evaluating telemetry and entries in the patient's Electronic Medical Record. It can also flag possible inappropriate entries when they are entered into the patient's EMR.

- **Mobile communication devices**   Text messaging devices and cell phones can be carried by nurses during their shift so that they can monitor all their patients as they go about their duties. For example, telemetry can be transmitted to the nurses' station's monitor and made available to the nurse's mobile communication device. Likewise, alerts, along with the patient's condition, can be transmitted by analytical computer software to these mobile communication devices.

- **Electronic charting**   Electronic charting computer software makes available a clearly written copy of the patient's charts to the patient's healthcare team. Charts can be used simultaneously, so there is no down time waiting for the chart to become available.

- **Electronic charting devices**   Electronic charting devices are small, mobile computers available at bedside to view and update the patient's EMR. Information about the patient that is entered into their chart at bedside is immediately available to other members of the patient's healthcare team—even from offsite locations, such as the physician's office, or by using the physician's mobile communication device.

- **Pharmaceutical tracking system**   A pharmaceutical tracking system eliminates or reduces time wasted by nurses waiting for delivery of medication from the pharmacy. The pharmacy reads electronic charting software that generates a list of medication for each patient and the time it is scheduled to be administered. The bar code on each medication is scanned into the pharmaceutical tracking system when it leaves the pharmacy and when the medication is placed in the patient's drawer in the medication cart. The computerized pharmaceutical tracking system automatically alerts the pharmacy, via an e-mail alert, a half-hour before an undelivered medication, that missed the routine delivery, is due to be administered. This gives the pharmacy time to get the medication to the unit before it is due to be administered.

- **Scheduling computer software**   Scheduling computer software enables the Nurse Manager to maintain a skills inventory for each staff member, and match qualified staff members with tasks that are to be performed during the shift. It alerts the Nurse Manager when a staff member is overallocated—assigned to perform two or more tasks at the same time. The software also generates an assignment list for each staff member.

Managing Scarce Resources Through Better Organization

A Nurse Manager can maximize the efficiency of staff resources by organizing staff members so that scarce resources focus full attention on their primary responsibilities. As a Nurse Manager, your staff should be organized so that tasks that are not required to be performed by scarce resources are delegated to appropriate support personnel.

Before you begin to reorganize your unit, you need to create three lists for your unit: a task list, a staff positions list, and a personnel list. The task list contains your unit's needs. The staff positions list identifies human resources required to fulfill those needs. The personnel list specifies which healthcare providers are available to fill staff positions.

Start by creating the task list. Write down all tasks that are routinely performed during each shift. Alongside each task, identify the following:

- Duration of the task

- Skill set required to perform the task

- Restrictions on who can perform the task

- Times during the shift when they are performed, if known

Create the staff positions list next. Write down each approved position for your unit, regardless of whether the position is currently filled. Alongside each position, write the skill set required for the position. Include requirements and restrictions set by law, hospital policy, and bargaining agreements.

Finally, create the personnel list by writing down each person on your staff. List each person's skill set and any requirements or restrictions related to their employment, such as availability.

Comparing skill sets of authorized positions with skill sets required for each task enables you to determine if you have sufficient staff positions to meet your needs. And by comparing the skill sets of authorized positions to the skill sets of your current staff, you'll know which resources are scarce in your unit.

REORGANIZING YOUR STAFF

As a Nurse Manager, one of the most effective ways to manage your staff is to organize them into patient care teams. Each team has a leader—the RN—and is assigned a group of patients. The team leader is responsible for care of their patients, and the team is responsible for completing items on their task list for their patients.

Each team should operate using the *management-by-exception theory*, which requires each team member to perform routine tasks with little or no supervision from the team leader, once the team member has won the team leader's confidence. The team leader should become involved only when there is an exception to the routine. This frees the team leader to perform tasks that only they are qualified to perform. For example, the team leader shouldn't be consulted when the nursing assistant takes the patient's vital signs and performs bedside care unless something unusual happens, such as the patient spiking a fever or the discovery of a bedsore.

Centralize administration and support functions to avoid patient care teams fighting over scarce resources. Staff scheduling, coordinating with other departments, and other administrative duties should be performed by a centralized team, and not by each team leader. Staff scheduling should be handled by the Nurse Manager. The unit secretary should coordinate activities with other departments, such as tracking down labs and medications.

Support functions that cross over teams should operate as their own team. Let's say there is an insufficient number of nursing assistants. Rather than assigning them to patient care teams, consider having them organized into their own team with their own team leader who isn't an RN.

Leaders of patient care teams make requests for support to the appropriate centralized team. The leader of the centralized team then assigns team members to fulfill each request. Suppose there are two patient care teams, Team A and Team B. The leader of Team A submits a request to the nursing assistants team to have Team A's patients' evening care completed by 10 P.M. Once the request is submitted, Team A's leader can focus on performing their own tasks. The leader of Team B makes a similar request for Team B's patients. The leader of the nursing assistants team creates a schedule for team members so that both requirements can be fulfilled within the deadline set by the patient care team leaders. Only if there is an exception (e.g. inability to meet a deadline, something unusual occurs with the patient, etc.) is the patient care team leader contacted.

Avoid the following common pitfalls of organizing a staff into teams:

- **Competition**   Teams are not in competition. Extinguish any sign of competition. Try to foster the attitude of "when one does well, we all do well."

- **Accountability**   Measure success on a unit basis and not on a team basis. If one patient in the unit isn't cared for properly, then the entire staff fails, and not just members of that patient's care team.

- **Communication**   Open communication must be the rule. A team member can speak with anyone, including making requests for support from a central team, to ensure that the team member's patient receives proper care.

- **Focus**   Keep everyone in the unit focused on providing quality patient care, and not simply completing items on a task list.

- **Empowerment**   Each staff member is authorized to make appropriate routine decisions related to patient care within the scope of their license and responsibility. Nobody needs approval to do the right thing.

Strategies for Managing Scarce Nonhuman Resources

Nonhuman resources are money, supplies, and equipment that the unit uses to deliver patient care. With a decline in reimbursements from third-party payers, hospital administrators impose strict financial controls on each department, resulting in reduced funding for supplies and material. It is therefore important that the Nurse Manager take steps to manage nonhuman resources. Here's what you can do as a Nurse Manager:

- **Charge back** The cost of supplies used by your unit is charged back to your unit's budget each month. It is critical that you review these charges carefully to be sure that supplies charged to your unit were actually used by your unit. Doing so will help preserve your scarce financial resources.

- **Restocking supplies** Explain to your staff how they can reduce unnecessary cost by not overstocking supplies in a patient's room, as once the supply is in one patient's room it cannot be returned to the general floor stock.

- **Proper accounting** Instill in your staff the importance of charging the patient's account for supplies used to care for the patient. Supplies that are used but not charged back to a patient become unaccountable and are usually charged to the unit.

- **Hording** Avoid hording supplies that are perceived to be in short supply, because hording exacerbates the problem by disrupting the normal use of supplies.

- **Share resources** Develop a plan to share scarce resources with other departments before the scarcity is realized. In this way, tasks requiring the scarce resource can be scheduled across departments to ensure that the resource is available when it is needed.

- **Owning versus renting** Evaluate the difference in cost between owning or renting equipment. Renting the equipment might be less costly in the long run than owning, because you pay only when the equipment is needed by a patient, and you avoid ancillary expenses such as maintenance.

- **Cost effectiveness** Analyze if new technology will actually save time and money over current technology.

Summary

Nursing management can be a stressful yet rewarding career choice. As a nurse, you have already learned to manage groups of patients and resources. As a Nurse Manager, you will perform those tasks on a larger scale with added responsibilities. This book is designed to give you the skills and resources to be successful as a Nurse Manager. Use it as a foundation on which to build and enhance your career, as well as your personal satisfaction with your ever-adverse, and always wonderful profession.

A resource is something needed to complete a task, and a scarce resource is a resource that has limited availability because there is more demand for the resource than there is availability of the resource. The most important scarce resource is a budget.

The staff should have a clear understanding of how the healthcare facility is—and is not—reimbursed for expenses related to patient care. Explain to your staff the difference between the price the hospital charges patients and third-party payers, and the actual cost incurred to care for the patient. The price covers patient care plus overhead and an adjustment to reflect reduced prices paid by healthcare plans.

A preliminary budget finances primary and secondary needs of your patients, your staff, your unit, and yourself. A primary need is a resource that is required to achieve your primary goal. A secondary need is a resource that indirectly supports a primary need.

Develop a strong, unshakable rationale for each item in your preliminary budget and relate it to patient care and reimbursements. Campaign to get your preliminary budget adopted. Your power to influence how scarce resources are allocated comes from your position as a manager within the organization and from your personal power.

Second only to scarce financial resources is the shortage of RNs. There are not enough nursing instructors today to teach all those who want to join the nursing profession—and there will be even fewer instructors in the near future.

Manage scarce nurse resources by using technology to reduce wasted time and focus on care that can only be given by RNs. Maximize the efficiency of staff resources by organizing your staff so that scarce resources focus full attention on their primary responsibilities. Organize your staff so that tasks that are not required to be performed by an RN are delegated to appropriate support personnel.

Quiz

1. What helps reduce unscheduled visits to the patient's room by a nurse?

   (a) Locating the nurses' station close to patient rooms

   (b) Telemetry

(c) Electronic charting

(d) None of the above

2. The accounting method used to allocate the cost of supplies within the hospital is called?

(a) Charge back

(b) Reimbursement

(c) Claims management

(d) None of the above

3. The most scarce resource is:

(a) Beds

(b) Money

(c) Nursing assistants

(d) None of the above

4. Personal power comes from your personality traits.

(a) True

(b) False

5. Electronic Medical Records can be used simultaneously by several caregivers so there is no down time waiting for the chart to become available.

(a) True

(b) False

6. What practice can exacerbate availability of scarce resources?

(a) Hording

(b) Sharing resources

(c) Chargeback accounting method

(d) None of the above

7. Sometimes, renting equipment is more cost effective than purchasing equipment.

(a) True

(b) False

8. Which of the following can be used to support the rationale for your preliminary budget?

(a) Benchmarking other hospitals

(b) Lowering exposure risks that lead to litigation

(c) Sharing resources with other departments

(d) All of the above

9. The price charged to patients by the hospital covers patient care plus overhead and an adjustment to reflect reduced prices charged to third-party payers.

(a) True

(b) False

10. The goal of lobbying for funding is to champion your unit and not minimize the needs of other units in the hospital.

(a) True

(b) False

ANSWERS

Chapter 1

1. b. Joint Commission on Accreditation of Healthcare Organizations
2. a. True
3. d. All of the above.
4. d. Systems that break down at various points.
5. a. True
6. d. None of the above.
7. b. False
8. d. All of the above.
9. b. False
10. a. True

Chapter 2

1. a. Peter Drucker
2. a. Behavioral theory
3. c. Guide and influence others without necessarily having the authority to direct their behavior
4. a. True
5. c. The Nurse Manager makes all the decisions.
6. d. Transactional theory
7. a. True

8. b. Contingency theory

9. a. True

10. a. True

Chapter 3

1. b. Case model

2. a. Place the registered nurse back at the patient's bedside

3. c. Moves the patient through the healthcare system in the most cost-efficient manner without sacrificing the quality of patient care

4. a. True

5. a. True

6. a. The level of care required by each patient

7. b. False

8. c. A complaint that management violated the collective bargaining agreement

9. b. False

10. a. True

Chapter 4

1. b. Registered nurse

2. d. All of the above

3. d. All of the above

4. b. False

5. a. True

6. a. Assess whether the nurse supervisor's decision is reasonable and within the scope of his or her role

7. a. True

8. b. Legally responsible if responsibility for patient care is delegated to an incompetent staff member

9. a. True

10. a. True

Chapter 5

1. b. When the Nurse Manager identifies and verifies facts related to a conflict
2. c. Resolve the conflict sufficiently to provide quality patient care
3. c. Define a process used by a person when adopting a new concept
4. a. True
5. a. True
6. d. All of the above
7. a. True
8. d. All of the above
9. a. True
10. a. True

Chapter 6

1. c. An appeal of disciplinary action against a member of a collective bargaining unit
2. a. A supplement to a bylaw
3. d. All of the above
4. b. False
5. b. False
6. b. Bylaws
7. a. True
8. d. All of the above
9. b. False
10. a. True

Chapter 7

1. b. Summons
2. a. Criminal law
3. b. Malpractice
4. b. False
5. b. False

6. c. The nurse proceeds with treatment after learning that the patient doesn't fully understand the risks associated with the treatment after the patient signs the informed consent

7. b. False

8. a. Carries out an incorrect medical order

9. a. True

10. b. False

Chapter 8

1. b. Economy

2. a. Free market

3. c. Consumer-paid services

4. b. False

5. b. False

6. a. Patients who have medical coverage

7. a. True

8. a. Cost shift

9. a. True

10. a. True

Chapter 9

1. b. At its breakeven point

2. a. Capital budget

3. b. Details the organization's long-range plans

4. a. True

5. a. True

6. c. Cash flow

7. a. True

8. a. Any consecutive 12 months

9. a. True

10. a. True

Chapter 10

1. c. There is a ten-day cooling-off period
2. a. A claim that the collective bargaining agreement was violated
3. b. Binding arbitration
4. a. True
5. a. True
6. d. All of the above
7. a. True
8. b. Must be approved by union members and the hospital's board of trustees
9. a. True
10. a. True

Chapter 11

1. a. The facts of the activity
2. a. Is an easily recognizable accomplishment
3. b. A part of the activity turned over to the person who requested the activity
4. a. True
5. a. True
6. d. All of the above
7. a. True
8. d. All of the above
9. a. True
10. a. True

Chapter 12

1. a. Informatics
2. c. A dedicated computer
3. a. A commercial application program
4. a. True
5. a. True
6. b. Electronic Medical Record

7. a. True

8. b. Telehealth

9. a. True

10. a. True

Chapter 13

1. d. All of the above

2. d. All of the above

3. d. All of the above

4. b. False

5. a. True

6. d. All of the above

7. b. False

8. d. All of the above

9. a. True

10. a. True

Chapter 14

1. b. Telemetry

2. a. Charge back

3. b. Money

4. a. True

5. a. True

6. a. Hording

7. a. True

8. d. All of the above

9. a. True

10. a. True

INDEX